Simple K-5 Science Investigations

Solutions for doing science in the classroom

Christopher P. Garside

Seven Sides Publishing

Seven Sides Publishing has a mission to improve the teaching and understanding of science. To contact us, send an email to simpleinvestigations@sevensidespublishing.com or visit our website at sevensidespublishing.com.

ISBN: 9798836816926

Published by: Seven Sides Publishing, Cypress, TX.

Table of Contents

Introduction

To help teachers teach science through investigations, Seven Sides Publishing has provided a series of lab manuals for Elementary Science, Middle School Science, as well as Biology, Chemistry, Physics, Environmental Systems, and Earth & Space Science. These manuals are a rich resource for structure and investigations. There is a shortage of user-friendly labs that easily allow teachers and students to perform investigations quickly. Too many labs have too much busy writing within them, where teachers do not want to take the time to read everything to figure out if it would be good for them to use with their students. If the teachers do not want to read it, do you think the students do? So we have taken many of the traditional labs that have been around for decades and simplified them, making them easy to read and perform. We have also added some new original labs that have never been seen before. There have been efforts to have teachers do more investigations with their students, but there is no plan or solution to deal with the real issues teachers have in preparing to do this. The book How to Teach Science Through Investigations has the plan, and the Simple Investigations Lab manuals have the solutions so students can learn science through investigations with minimal effort. Teaching science through investigations will make your classrooms more efficient, where students learn content and practice skills simultaneously. Science is a process of doing. Doing this process is the most efficient way for students to learn science and be able to use it in the future. We live in a culture where science-literate people are needed for jobs, but too few can be found. If you incorporate these investigations with virtual labs (that I will point you to in each section of the lab manual) and skill/math practice, you will not need to fill in gaps by giving lectures. All content can be learned through investigations and practice. Remember, we only remember 5-20% of what we hear. That 20% is when you are really interested in the content. But hearing practices no science process skills and does not activate any higher cognitive thought. Lecturing is not a good option. We remember 75-80% of what we do/experience and 90-95% of what we teach. Investigations allow us to keep our students in these higher retention percentages. Teaching through investigations also works because students spend more time in class at higher Bloom's Taxonomy levels, staying in zones C and D on the Rigor Relevance Chart when they perform investigations. If you add the physical way students are stimulated with the hands-on experience, you cannot deny the level of learning will be much higher while students perform investigations. This manual gives you the resources you need to teach Kindergarten through 5th Grade Science through investigations.

We separated each of these sections in the manual by grade level. We will follow the Kindergarten through 5th Grade Science Texas Essential Knowledges and Skills and Next Generation Science Standards to make it easy to find the labs you want and need for your classes. At the beginning of each lab, we put the materials you will need in boldface in the directions; this saves time for your lab preparation. Some investigations are shown again (repeated) in different grades because of differences in the TEKS and NGSS. We wanted them to be easy for teachers to find no matter which standard they follow.

Because they are so simple, these labs can be modified to fit whatever equipment you have. There are very few labs that I have used in my career that I did not alter how I presented them. One reason we wrote these labs this way is to customize them to the Texas Essential Knowledge and Skills and Next Generation Science Standards. We also wrote them how we thought a teacher would want to use them.

Virtual Labs

Hands-on labs are not the only way for students to learn science, but they are the most effective. However, virtual labs can be used with these hands-on labs. Many investigations physically cannot be done hands-on, so some experiments will have to be done virtually. There are three sources that I have used in the past that have a good number of resources. **Physicsclassroom.com** and **PhET.colorado.edu** are free to everyone and are great to use. **Physicsclassroom.com** has teacher notes and activities/exercises that guide students through Physics and Chemistry Interactives. You can find them under the simulation and open, download, or print the PDF. **PhET.colorado.edu** has a variety of activities of different levels that you can explore to go through their simulations. They are also easy to download and print. **ExploreLearning.com** is expensive, but the quality of its product is much higher than the other two. When you click on a Gizmo, you can also click on lessons and find the Student Explorations that go with each Gizmo that you can modify, download, and print. They are written at a very high quality, making the students think like a scientist. At the end of each section (except Kindergarten) of this lab manual, we include a list of virtual labs from these organizations that would be great to use with investigations from that grade level. Please remember virtual labs should never replace hands-on labs. If the students can learn the content live, that should be the priority because it is more of an experience that will be remembered. There are many other virtual simulations out there, but none so far have moved me to use them over the three I have mentioned here.

Kindergarten Science Investigations

The Kindergarten investigations are meant to be done less independently, with the teachers asking the students questions orally since most kindergartners are just starting to learn to read and write; this is why there is no space left after the questions. There is no virtual investigation list for Kindergarten also for literacy reasons. However, Grades 1-5 have lists of virtual investigations that could be used with the investigations in this manual.

Classifying Objects

Directions:

Have students find objects that fit the physical properties below. Have them draw or write the objects that fit each property:

Round	Rectangle	Hard	Soft	Light	Heavy

Red	Blue	Brown	Green	Yellow	Orange

Force and Motion Relay Race

Directions:

Read the directions and ask the students the questions

1) Place the **basketball** on the floor. Push the ball with a **broom** in a sweeping motion to cause the basketball to move faster. When does the basketball move?
2) Why does it move?
3) Place the **kid's rubber ball** on the floor and push this ball the same way with the broom you did with the basketball. When does the ball move?
4) Why does it move?
5) Which ball needed more force to move?
6) Which ball needed less force to move?
7) Get the basketball moving, then stop it with the broom. Do the same with the kid's rubber ball. Which ball was easier to stop?
8) Why was it easier to stop?
9) Which ball was harder to stop?
10) Why was it harder to stop?
11) Get the basketball and start moving it, then stop it and turn it in the opposite direction. Do the same for the kid's rubber ball. Which ball was easier to change direction?
12) Why was it easier to change direction?

Relay Race:

13) Make a course that changes direction several times (inside or outside) to push the different balls around in a race. Divide up into two equal teams, each with a broom. One team will push a basketball with a broom; one team will push the kid's rubber ball with a broom. Predict which team will finish the course first with all its team members.
14) Have the students do a relay race to move the balls through the course. Which team won?
15) Why were they able to win?

Forces in a Car Crash

Directions:

You will need a **toy car that winds up** when you pull it back and then moves forward when you let it go. You will also need a **penny** and a **rubber band**. Follow the directions and have the students answer the questions.

1) Place the penny on the car. Pull the car back to wind up the car. Let it move forward and have it crash into something.
2) What happens to the penny?
3) Why did the penny keep moving forward in the first crash?
4) Place a rubber band around the car and use it to fix the penny to the car. Repeat the procedure in # 1. What happens to the penny?
5) Why did the penny stay on the car for the second crash?
6) Explain the forces acting in the second crash.
7) What does the penny represent?
8) What does the rubber band represent?
9) How is this a good model to see the results of a car crash?
10) Why should we wear seat belts when we get in a car?

Inertia Lab Stations

Directions:

Station 1: Take a **ping pong ball** and use a **ping pong paddle** to bounce it off the door or wall a few times (you can go outside). Now take a **tennis ball** and use the paddle to bounce it off the door or wall a few times. Ask the students...

1) Which ball needs more force to hit it?
2) Why do you think that ball needs more force?
3) Which ball can you get to move faster off the paddle? Why?

Station 2: Roll a ping pong ball at a stationary tennis ball, and have them collide. Now roll the tennis ball at a stationary ping pong ball, and have them collide.

1) Which ball had more force and energy?
2) Why do you think it has more force and energy?

Magnets

Directions:

Give each pair of students two **bar magnets** and have them find out how they behave with each other and objects in the room. They should see that:

- They will attract to each other when opposite poles face each other
- They will repel each other when like poles face each other
- They will stick to things made of iron, cobalt, and nickel
- They will not stick to anything else

Ask the students:

1) How can magnets be made to pull on each other?
2) How can magnets be made to push each other?
3) Do you think a magnet will pull on the wall? Then have them test it.
4) Do you think a magnet will pull on the door? Then have them test it.
5) Do you think a magnet will attract an **aluminum can** or **aluminum foil**? Then have them test it.

Have the students go around the room and find things their magnet will pull on. Then, as a class, fill in a chart showing things that magnets pull on and don't pull on in the room.

Pull →←	Not Pull

6) What do all objects with magnetic pulls have in common?
7) What is the only thing a magnet will push away?

Balls in Motion

Directions:

Get a segment of a **Hot Wheels track** and a **small ball**. Place a **small sticker** on one end of the ramp to mark where you will place your ball for each trial to let it roll down the ramp; this keeps the distance your ball will be accelerating down the ramp constant. Place some **books** under one side of the track to raise that end to create a ramp. Draw Data Table 1 on the front board of the room. Follow the directions and ask the students these questions during the investigation.

1) Set the ramp so the ramp's bottom is along the edge of a tile on the floor. Most tiles in schools are 1 foot in length. Clear a path for 10 feet.
2) Adjust the ramp height so the ball will go a short distance. Have a student measure the height of the ramp with a ruler. Write that height in Data Table 1.
3) Place your ball on the ramp and let it roll down (do not push). How many squares did the ball travel?
4) How do you think we can get the ball to move farther?
5) Raise the ramp's height so the ball will go a longer distance. Have a student measure the height of the ramp with a ruler. Write that height in Data Table 1.
6) Place the ball on the ramp and let it roll down. How many squares did the ball travel?
7) Why do you think the ball traveled farther this time?

Data Table 1

Trial	Height of Ramp	Number of Squares the ball rolled
1		
2		

8) Describe the motion of the ball as it moved down the ramp. If you have to, place the ball on the ramp and let it go to observe it move down the ramp.
9) What force caused the ball to speed up on the ramp?
10) How could we change the ramp to make the ball have a higher velocity at the bottom of the ramp?
11) How are ramps used to help us make life easier?

Observing Changes in Motion

Directions:

You will need a **ball** to throw and bounce off objects. You will need a safe place to allow the ball to bounce in different directions. Gently throw your ball up in the air.

1) How did your ball move?

2) What force caused the ball to move up?

3) Why do you think it changed direction in the air?

4) As the ball moves up, how does its motion change?

5) When the ball comes down, how does its motion change?

6) What force is causing the change in motion?

Gently throw your ball to bounce it off a wall.

1) What force caused the ball to move towards the wall?

2) When the ball hit the wall, how did the ball change motion?

3) What forces caused the change the motion?

4) How did you see the ball change its motion today?

5) What force is always acting on a ball?

The Force of Collisions

Directions:

Place a piece of **masking tape** on the floor and line up a **ramp/track** up to it. Place the **softball** at the bottom of the track on the tape. Draw Data Table 1 on the board.

1) Roll a **ping pong ball** down the track and measure how far it moved the softball with a **meter stick**. Record this in Data Table 1 on the board.
2) Roll a **tennis ball** down the ramp and measure how far it moves the softball. Record this in Data Table 1 on the board.
3) Roll a **baseball** down the ramp and measure how far it moves the softball. Record this on the data table on the board.

Data Table 1

Ball	Mass (g)	How far it moved the softball (cm)	Force Rank
Ping Pong	3g		
Tennis	58g		
Baseball	145g		

1) Which ball moved the softball the farthest?

2) This ball had the greatest force. Rank the force of the balls at the bottom of the ramp from highest to lowest in Data Table 1.

3) In this experiment, what caused the biggest change in force?

4) How else could we change the force?

5) Test it to see if it works.

I See Light Energy

Directions:

Ask the students the questions below and follow the instructions:

1) Do you think you can see your hand if no light shines on it?
2) The lights are on; can you see your hand now?
3) What color is it?

Turn off the lights.

4) Can you see your hand?
5) Does it have the same color it had when the lights were on?

If you are not in an interior room that blocks out most if not all light, find a place to take your students where they cannot see their hands in front of their face. Turn off the light for a couple of seconds, but not too long, so they do not get scared. After turning off the lights when they are looking at their hand and turning them back on, ask them...

6) Could you see your hand?
7) Why do you think you could not see your hand?
8) Did your hand have the same color as when the lights were on?
9) What do you think allows you to see objects?
10) What do you think allows you to see colors?

Tell the students: Light allows energy to move through space.

11) What do you think that energy can do? Answers may be: to allow us to see things, allow plants to grow, communicate signals, and see colors.

If you have a **radiometer** (which can be found in Hobby Lobby), put it up in the window and have the students watch it when it is sunny and then cloudy. Then sometime later, ask...

12) Why do you think the radiometer changes how it spins?
13) Which side, the black or white, is the light absorbed, and which side is it bouncing off? Black absorbs, adding energy, causing it to spin. The white cause the light to bounce off, reflecting energy away.

Traveling Light

Directions:

Have students look through a window. Ask them...

1) What do you see?

2) How is this different than looking through a wall?

3) Which object bocks light (a window or a wall)?

4) Which object allow light to travel through (a window or a wall)?

Turn on a **lamp** in the room with (no shade) and turn off all the other lights. Ask the students...

5) Where is your shadow?

6) Why is your shadow there?

7) Is your body allowing the light to pass through it or blocking it?

Have students hold some objects up to the light and see if they cast a shadow. Shadows will tell us if the object will block light or allow it to pass through. The darker the shadow, the more the object blocks light; the lighter the shadow, the less the object blocks light. Draw Data Table 1 on the board, leaving the examples below out. Fill it in with your students and tell if objects will let most light pass through, block some light, or block all light. Some examples are given below.

Data Table 1

Allow most light to pass	Block some light & allow some light to pass	Block all light
Window	*Sunglasses*	*Walls*
Air	*Colored glass or plastic*	*Human body*
Glass/ plastic cup	*Frosted glass*	*Mirrors*
Plastic wrap	*Shower curtain*	*Frames of glasses*

Blocking the Heat of Sunlight

Directions:

Go outside where there is pavement that is in the sun and pavement that is in the shade. Have the students feel how warm the pavement is in the shade (it is best to use the back of the hand since it is more sensitive), then feel how warm the pavement is in the sun. Ask the students...

1) Which was warmer?

2) Why do you think that pavement was warmer?

3) Does sunlight cause objects to heat up?

4) Do you think we can cool the hot pavement? How?

Have the students develop ideas to help keep the pavement from getting hot on the playground. As a class, design a way to do this with your chosen materials (**tarp**, **tent**, **umbrella**, or a **board** resting on **chairs**). Set it up in the sun. Let it sit in the sun for an hour and come back and see if the pavement is cooler under the shade than in the light.

5) Why did the pavement cool down?

6) What could we do to make the pavement even cooler?

7) How is this information important to help make our lives easier?

Contest: Divide the class into groups and have them design and build an object to cool a spot on the concrete the most. Use thermometers to see which design causes the biggest temperature drop in the same amount of time. Groups cannot use ice or any electronic devices to cool the surface in any way.

Night and Day

Directions:

Ask students the questions below. I have added some possible answers and information you may want to communicate with the students.

1) What do you see in the sky during the day? *Sun, clouds, birds, planes, and sometimes the moon.*

2) What do you see in the sky at night? *Moon, stars, lights of planes, planets. The brightest stars are usually planets like Venus, Mars, Jupiter, and Saturn.*

3) Have you ever seen the moon during the day? *It spends the same amount of time in our sky at night as it does during the day.*

4) Do you see the moon every night? *No, it also changes its shape every night.*

5) Why do you think you see the moon more at night than during the day? *The moon is not as bright as the sun and reflects the sun's light. There is less contrast during the day than at night, making it easier to see at night.*

6) What object provides light for us to see during the day? *Sun*

7) What objects provide light for us to see at night? *Stars, moon, electric lights, and fire*

8) Is the moon making light or reflecting light? *Reflecting*

9) Where is the moon reflecting light from? *Sun*

10) When is it warmer (day or night)? *Day*

11) When is it colder (day or night)? *Night*

12) Why is the day warmer than the night? *Because the sun is sending energy to the Earth's surface as light during the day. Some of the energy goes out into space at night.*

13) Does the sun set at the same time every night? *No*

Patterns with Sunrise and Sunset

Directions:

Have students find out when the sun rises and sets each day and chart it together as a class. Chart the high temperature and low temperature each day. This chart will be useful in future lessons. A sample of a chart is below. You can modify this activity using an almanac to get this data from previous years.

Date	Time Sunrise	Time Sunset	High Temperature	Low Temperature

My Rock Collection

Directions:

Have students gather **rocks** from home and outside (no valuable rocks from home). Have them classify them in the categories in the charts below. Then count how many rocks they have in each category.

Under 1 inch	Between 1-2 inches	Over 2 inches

Smooth	Rough

Light color	Dark color

Round	Skinny

Shiny	Dull

You can take your totals from each student and make charts for the whole class, adding each student's data to the class data tables.

1) Which category had the most rocks? Why do you think that was?

2) Which category had the least amount of rocks? Why do you think that was?

3) What do you think causes the rocks to be different colors? *The minerals in it.*

4) Were there any rocks that allowed light to travel through? *Some of these would be called crystals.*

Weather Changes

Directions:

Ask students…

1) What is the source of heat for the surface of the Earth? *Sun*
2) How does the heat/thermal energy get here? *Sunlight*
3) What do you think causes days to be hotter? *Longer days, shorter nights, sunny days*
4) What do you think causes days to be cooler? *Shorter days, longer nights, cloudy days*
5) Why does the weather change? *Air masses are constantly moving over the surface of the Earth from its rotation*
6) How do you think a cloudy night differs from a night with clear skies? *Clouds trap heat, causing it to stay warmer like a blanket. Clear skies allow heat to leave the Earth.*
7) Which season is the warmest? *Summer*
8) Which season is the coldest? *Winter*
9) Why do you think some seasons are warmer than others? *The amount of light hitting the surface.*

Look at the data from the activity: Patterns with Sunrise and Sunset

10) Which day had the earliest sunrise and latest sunset from our data?
11) Was it warmer or colder then?
12) Why do you think that happened?
13) Were the days getting longer or shorter?
14) Were the days getting warmer or colder over time?
15) Which season has the longest days? *Summer*
16) Which season has the shortest days? *Winter*
17) Does the length of the day have anything to do with how warm it will be? Why? *Yes, more light, more heat.*
18) When do the days have equal daytime and nighttime? *Spring and fall (equinox)*
19) What would happen if one side of the Earth stayed day and the other stayed night all the time? *The dayside would get very hot and the night side would get very cold. Most life on Earth would die.*

Keep taking data on the sunrise and sunset, tracking the high and low temperatures. Draw attention to the data when the season changes and ask…

20) Are the days longer or shorter?
21) How should the temperature change during this season?

The Wind

Directions:

Give each student a section of a **party streamer** and go outside. Have the students hold them up and allow the wind to move them. Ask the students…

1) Why is the streamer moving? *Particles are hitting them, forcing them to move*
2) What is the air made of? *Molecules, atoms, or particles*
3) How do we know when air moves? *Objects are pushed by it*
4) When and where can we find wind? *Currents of air, from different pressures and different temperatures*
5) What is wind? *A collective movement of particles*

Take out a **pinwheel** and ask students…

6) How do you think we could measure how fast the wind is moving? *How fast it is spinning*
7) How can we tell which direction the wind moves? *The direction it is facing when moving the fastest*
8) How can we tell the wind is a force? *It moves objects*
9) Is the force of wind a push or a pull? *Mostly a push but can be a pull.*
10) How do we detect wind as a push? *Push on us*
11) How can we see the wind is a pull? *Driving and things fly out of the back of a truck, or a tornado lifts the roof off of buildings.*

Go back inside and have the students place their party streamer on their desk/table and blow over the top of it.

12) What does your streamer do? *They should all lift in the air.*
13) Was this a push or a pull? *Pull*

If the students ask why this happens, tell them as a gas or a liquid (both are fluids) move faster, their pressure decreases (Bernoulli's Principle). It causes a curveball to curve, a golfball to slice or hook, and an airplane wing to lift into the air.

Rocks Soil and Water

Directions:

Have a **bottle of water** in front of the students and ask...

1) What do we use water for? *Cleaning, drinking, growing food and plants, cooking, baking, processing materials, mixing cement, giving to their pets...*

Have a **cup of soil** in front of the students and ask...

2) What do we use soil for? *Growing plants, leveling our lawns, burying things, growing food, building athletic fields and golf courses...*

Have some **stones/rocks** in front of the students and ask...

3) What do we use rocks for? *Build walls, gravel roads and paths, weapons, build buildings in the old days, line flower beds and gardens to trap water in the ground...*
4) What is something we can do with all three? *Hold in moisture for plants*

Get some **dry beans** and have the students push 2 dry beans into their cup of soil and place some small stones over the top, leaving space for the plant to push through. The stones will help hold in the moisture. *Have the teacher take a cup, plant two seeds, water it (label it H_2O), cover the top with stones with no space so the plant cannot grow through, and places it in the light.*

- Have half the class pour water on their seeds and the other half not pour water on their seeds. The ones that were watered have the students write H_2O in the cups.
- Take half the cups that have been watered and half that have not been watered and place them in a dark cabinet where they will get no light.
- Have the rest of the planted cups say out in the light. If you have a window sill, you can place them there.
- Have the students check the plants in the light to see how they grow. Do not open the cabinet to the plants in the dark.
- Have the students watch to see which ones grow taller, the ones that were watered or the ones that were not watered.

Once the plants have grown a few inches tall, do the next lab: Give Plants Light or Give them Death.

Give Plants Light or Give them Death

Directions:

Have the students look at the plants that grew in the light and then bring out the plants in the dark and place them next to the plants that grew in the light. Ask the students...

1) Which plants grew the biggest? *The ones with water will have grown the most. The ones in the dark and watered should be longer than those in the light, but they should be pail and look unhealthy. Light inhibits growth but allows the energy from light to feed the plant helping it grow healthy. If a plant goes without light for too long, it will die.*
2) Which plants look the healthiest? *The plants that were in the light and watered.*
3) What do we know plants need to be healthy? *Soil, water, and light.*

Show the cup that the teacher planted with rocks covering the top with no place for the plant to poke through. Ask the students...

4) Did the plant grow when I covered it with rocks?
5) What else do you think plants need to grow? *Space, if they do not have it, plants can't grow any bigger.*
6) What do you think the plants used to make food from the light? *Water, air (carbon dioxide), and nutrients from the soil.*

Have the students look at the healthy plants. Have the students draw a picture of the plant's roots, stems, and leaves and label it. Have the students point to the parts of the plant they drew as they answer the questions.

7) Which part of the plant takes in the water? *Roots*
8) Which part of the plant takes in the nutrients from the soil? *Roots*
9) How do we know roots take in water and nutrients from the soil? *Because they are there in the soil with the water and nutrients.*
10) Which part of the plant takes in the light? *Leaves*
11) Which part of the plant takes in the air (carbon dioxide)? *Leaves, they have tiny holes to bring in air.*
12) How do we know the leaves take in light and air? *Because they are above ground in the air.*
13) Which part holds the leaves up to the light? *Stem*
14) Which part transports the nutrients and water up to the leaves? *Stem*
15) How do we know? *Because it is between the roots and leaves. They act like straws.*

Can We Keep it? *(No)*

Directions:

Catch a little **lizard, gecko**, or another **small animal** that is safe to catch temporarily and place it in a **jar** with small holes in the **lid** or a **terrarium**. Bring it to your class and say, "Look what I found..." ask the students...

1) What do we need if we want to keep it alive? Keep asking, "What else would we need?" until the students have mentioned all: *water, a source of fresh air, and food. When in the wild, they need space to be able to find enough food; they need a shelter to hide from bad weather and predators.*
2) Why are there slits in the cage/holes in the jar's lid? *So they get air*
3) What would happen if it did not get fresh air? *It will Die*
4) What kind of food do you think it needs? *Lizards only eat live bugs. They need lots of space to catch live bugs.*
5) What will happen if it does not eat the food we give it? *Die*
6) How will we feed it every day?
7) How will it get water?
8) Will the chemicals in the tap water kill it? *We don't know.*
9) Would it be better for the lizard to keep it with us or let it go? *Let it go.*
10) Why? *Have them list that it needs to catch live food, needs lots of space to find enough food, and they need a source of water. It probably already has a shelter it likes to hide in.*

A Class Pet

Directions:

Get a class **pet** and **materials to care for it** that are appropriate for your class. Research and find one you and your class are willing and able to care for, like a fish, bird, hamster, gerbil, or Gini pig. Assign students to monitor its food and water each day. Have a procedure where students safely clean the cage periodically, so it stays healthy and does not stink. You can use some of the questions above to discuss its need for air (fish get oxygen gas from the water with gills), water, food, space, and shelter with the class. They will be reminded of these needs every time someone takes care of it.

Flower to Fruit

Directions:

Bring enough **potted flowers** into class for the students to see, draw, and label. Lay some **paper towels** on a table or counter and carefully pull the flowering plant out of the pot so students can see the roots. Give the students a **magnifying glass** to explore the plant.

- Have the students draw the plant using the magnifying glass to look at the root, stem, leaves, and flowers. Have the students use the magnifying glass to see the details of the plant to help them draw and label the roots, stems, leaves, and flowers.
- Bring out a variety of **fruit with seeds** and tell the students that most flowers will eventually turn into fruit. Cut open the fruit to expose the seeds for the students to see. Have the students draw and label the fruit and seeds on the same paper as their plant and flower.
- Repot the flowering plant and clean up.

Grow a Garden

Directions:

Find a safe area for the students to plant **seeds** to grow a fruit (fruit is anything that contains seeds) and vegetable garden so the students can periodically watch the seeds turn into plants, and flowers, then produce fruit and vegetables. Build a border to keep the plants safe from the feet and toys of playing children.

Bring the students out and take pictures you will use later when the students draw a life cycle of one of your fruiting plants:

- To plant the garden
- When the plants germinate into a seedling, breaking through the ground. They will notice that different plants will grow at different rates.
- When the plant takes its adult shape
- When the plants flower
- When the fruit forms
- When it is time to harvest the fruit and vegetables, have the students wash and eat them, showing how we as human animals depend on plants for food.

You can take the seeds from your crop to later plant again with your class or future classes.

Plant Life Cycle

Directions:

Have the students draw the life cycle of one of the plants they watched grow in the garden.

1. Seed
2. Seedling
3. Plant
4. Flower
5. Fruit (which contains the seed to start again)

Make sure they draw each part of the life cycle after seeing it in the garden and label it while forming a circle so that the ending of one generation starts the next.

Example:

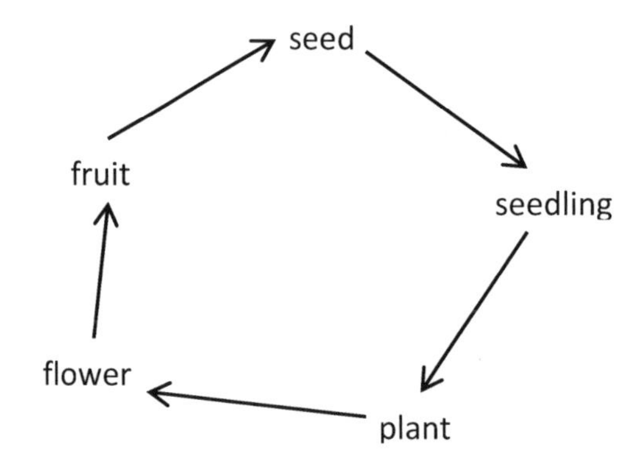

Question:

1) How do the baby plants resemble their parents?

Animal Parts

Directions:

Have the students look at your **class pet** and other animals from videos showing different animals interacting in their environment. Have students draw and label each animal's parts that allow them to...

- See
- Hear
- Move
- Grasp objects

Which type of soil do beans like best?

Directions:

Have the students plant a few **dry beans** in each type of soil in separate labeled **cups** or **pots**:

- **Sand**
- **Garden soil**
- **Loam**
- **Gravel**
- **Clay**
- **Soil from outside the building**

Make sure they all get equal access to light (set them on a window sill or under a grow light that you turn off at night), the same amount of water (measure it out with the same measuring tool), and they are in the same type of pot/cup.

Have the students observe when each sprouts, how they grow, and if any die. Then ask the students...

1) Which type of soil made the beans sprout first?
2) Why do you think this happened?
3) Which type of soil made the bean plants grow the tallest? Have the students measure with a ruler
4) Which type of soil made the bean plants grow the widest? Have the students measure with a ruler.
5) Which type of soil makes the bean plants seem the healthiest?
6) Why do you think this happened?
7) Which type(s) of soil had all the plants die?
8) Why do you think this happened?
9) Which type(s) of soil had some plants die?
10) Which type(s) of soil had no plants grow?
11) Why do you think this happened?

How Animals Change the Environment

Directions:

You may want to show students short videos where students see these organisms. Have the students answer the questions then they will draw a picture of an ecosystem with all the organisms.

1) How do squirrels change the environment to hide food?
2) How do beavers change the environment to have a house?
3) How do birds change the environment to lay eggs?
4) How does the groundhog change the environment to build a home?
5) Deer like leafy plants for food and need water to drink, so where do they live?
6) Fish have gills, so where do they live?
7) Where do we see plants growing?
    a. Why do plants grow there?

On a piece of paper, draw a picture of an ecosystem where they all can live. Then ask the students the next set of questions.

8) Humans need water, food, and shelter, so where can they live?
9) How do humans change the environment to have a home?
10) How do humans change the environment to have food?
11) How do humans change the environment to build things they want?

Draw a new picture of the same place, but now humans have moved in. Then answer the next set of questions comparing the two pictures.

12) How has the environment changed?
13) How did the land change?
14) How did the water change?
15) What happened to the plants when humans moved in?
16) Where did the animal's homes have to go?
17) Can humans and nature live together and have everything they want? Why?
18) How can humans reduce the impact on the other organisms so they all can live?
19) How was comparing these pictures useful in seeing the impact humans have on the environment?

You could have the students watch ***The Lorax*** (the old version) and see this play out. After watching the story, you can use many of the questions above to help a discussion.

Famous People in Science

Directions:

After students have performed some science investigations, have students watch videos or read to them stories about these famous people in science that have to do with what the students were investigating. Have them draw a picture of the person and describe what they are famous for.

Physics/Astronomy

Isaac Newton

Galileo Galilei

Nicolaus Copernicus

Johannes Kepler

Astronauts

Buzz Aldrin

Neil Armstrong

Mae Jemison

Sally Ride

John Glen Allen Shepard

Botanists

Carl Linnaeus

Agnes Arber

Ynes Mexia

George Washington Carver

Gregor Mendel

Kindergarten Science TEKS & NGSS Correlations

Classifying Objects Science, Kindergarten (b) 1EF 2B 5A 6

Force and Motion Relay Race Science, Kindergarten (b) 1ABCDE 2BD 3ABC 5ABCE 7; K-PS2-12

Forces in a Car Crash Science, Kindergarten (b) 1ABCDEG 2ABD 3ABC 4A 5AB 7; K-PS2-12, K-2-ETS1-2

Inertia Lab Stations Science, Kindergarten (b) 1ABCDE 2B 3ABC 5ABCE 7; K-PS2-12

Magnets Science, Kindergarten (b) 1ABCDE 2BC 3ABC 5ABDFG 7; K-PS2-12

Balls in Motion Science, Kindergarten (b) 1ABCDEF 2BCD 3ABC 4A 5ABCDEG 7; K-PS2-12

Observing Changes in Motion Science, Kindergarten (b) 1ABCDE 2B 3ABC 5ABE 7; K-PS2-12

The Force of Collisions Science, Kindergarten (b) 1ABCDEF 2BCD 3ABC 5ABCE 7; K-PS2-12

I See Light Energy Science, Kindergarten (b) 1ABCEG 2BC 3ABC 5ABCDEG 7 8A; K-PS3-1

Traveling Light Science, Kindergarten (b) 1ABCDEF 2BC 3ABC 4A 5ABCE 8AB; K-PS3-2, K-2-ETS1-1

Blocking the Heat of Sunlight Science, Kindergarten (b) 1ABCDEG 2ABCD 3ABC 4A 5ABCDEG 8B; K-PS3-12, K-2-ETS1-123

Night and Day Science, Kindergarten (b) 1ABEF 2BC 3ABC 5ABCEFG 9AB; K-ESS2-1

Patters with Sunrise and Sunset Science, Kindergarten (b) 1EF 2BC 3B 5ABCEG 9AB; K-ESS2-1

My Rock Collection Science, Kindergarten (b) 1ABCEF 2BC 3ABC 5ACE 10A

Weather Changes Science, Kindergarten (b) 1ABE 2BC 3ABC 5ABCDEFG 10B; K-ESS2-1

The Wind Science, Kindergarten (b) 1ABCDEG 2B 3ABC 4A 5ABCDE 7 10C; K-2-ETS1-2

Rocks Soil and Water Science, Kindergarten (b) 1ABCD 3BC 4A 5BEFG 11

Give Plants Light or give them Death Science, Kindergarten (b) 1ABCDEFG 3ABC 5ABCDEFG 12A 13A; K-LS1-1, K-ESS3-1

Can We Keep It? (No) Science, Kindergarten (b) 1AB 3BC 5BG 12B; K-LS1-1

A Class Pet Science, Kindergarten (b) 1ABCD 5BCDEG 12B; K-LS1-1

Flower to Fruit Science, Kindergarten (b) 1BCDFG 5DEF 12A 13A

Grow a Garden Science, Kindergarten (b) 1BCD 5DEG 12A

Plant Life Cycle Science, Kindergarten (b) 1BCDEF 3B 5DEG 12A 13ACD

Animal Parts Science, Kindergarten (b) 1BCEFG 5DF 13B; K-ESS3-1

Which type of soil do beans like best? Science, Kindergarten (b) 1ABCDEF 2BCD 3ABC 5ABCDEG 12A 13ACD; K-LS1-1, K-ESS3-1, K-2-ETS1-13

How Animals Change the Environment Science, Kindergarten (b) 1ABEFG 2AB 3ABC 4A 5ABDFG 12AB 13AB; K-LS1-1, K-ESS3-12

Famous People in Science Science, Kindergarten (b) 1AB 3AB 4AB

Kindergarten Equipment List for all Investigations

If you want to be able to do all the labs in this manual for Kindergarten, here is a list of all the equipment you will need (in order of appearance).

Basketball

Broom

Kid's rubber ball

Toy car that winds up

Pennies

Rubber bands

Ping pong balls

Ping pong paddles

Tennis balls

Bar magnets

Aluminum can/aluminum foil

Hot Wheels track

Small rubber balls

Small stickers

Masking tape

Softball

Baseball

Radiometer

Tarp/tent/umbrella/board

Lamp

Rocks/stones

Party streamers

Pinwheels

Bottles of water

Cups/tiny pots

Soil

Dry beans

Small live animal

Terrarium/glass jar with holes in the lid

Class pet and materials to take care of it

Potted flowers

Paper towels

Magnifying glass

Fruit with seeds

Sand

Garden soil

Gravel

Clay

Soil outside school

1st Grade Science Investigations

Classifying Objects

Directions:

Have students find objects that fit the physical properties below. Have them draw or write the objects that fit each property:

Sphere	Triangle	Rough	Smooth	Lighter than a chair	Heavier than a chair

Bigger than a soccer ball	Smaller than a soccer ball	Narrow	Wide	Light color	Dark color

Ice to Water and Back Again

Directions:

You will need **ice cubes** floating in **water** inside a large clear **beaker** positioned where all students can see. Then ask the students these questions...

1) What do you think ice is?

2) How does ice feel?

3) Is ice lighter or heavier than liquid water? Explain how you know.

4) What do you think will happen if we let it sit for a while?

5) What is appearing on the outside of the container that we must wipe off to see inside?

6) Where do you think it is coming from?

Let it sit and melt. After it melts, bring the student's attention back to it and ask...

1) Where did the ice go?

2) What is ice? How do we know?

3) How does water feel?

4) Can we change the water back into ice? If so, how?

Have students put their names on **paper cups**. Pour the water into their cups and have them place them in a **freezer**. Come back at the end of the day or the next day and see what happened to their water. Then ask the students...

1) What happened to the water?

2) How do we know this is ice?

    a. How does ice feel?

    b. Does this float in liquid water?

3) Can we change it back to liquid water? If so, how?

4) What else do you think can happen to water?

Pour water into a **beaker** and make a mark on the container where the water's surface is. Let it sit out for a few days and have the students see the water level drop. Then ask the students...

1) What do you think happened to the water?

2) Where did it go?

Changing the Rate of Evaporation

Directions:

You will need a **heat lamp** (can be an **incandescent light bulb** and a **work lamp** with an **aluminum shield**), four **glass/plastic beakers** (label them "A" "B" "C" "D") of the same size filled with the same amount of **water**, and two **plastic/glass Petri dishes** that can cover the openings of the beakers. Set two beakers under a heat lamp and two beakers away from the heat lamp. Cover one of the beakers with a Petri dish under the lamp and one away from the lamp. Build a data table like Data Table 1 below for the students to fill in. Then ask the students...

1) Do you think we can change how fast evaporation happens?

Show the students the setup of the investigation and then ask...

2) What do you think we will be testing with the water?

3) How do you think we will use the heat lamp?

4) Why do you think we are covering half of the beakers with Petri dishes?

5) Which beaker do you think will evaporate water the fastest?

Turn on the heat lamp and have students take turns finding the volume of each beaker each day, filling in Data Table 1. Make sure the lamp is not too close to the beakers to cause them to melt if they are plastic.

Data Table 1

Day	Beaker A	Beaker B	Beaker C	Beaker D
1				
2				
3				
4				
5				
6				
7				
8				
9				

Have the students use Data Table 1 and the lab setup to answer the following questions

6) Which beaker evaporated water the fastest? Explain why.

7) Which beaker evaporated the slowest? Explain why.

Is Cooking Reversible?

Directions:

You will need a **video camera** connected to a **smartboard** or **some other form of projection** to safely show the action of cooking taking place. You will want a **hotplate** or an **electric range** and a **skillet** to melt **butter**, cook an **egg**, and cook **popcorn**. Line up the camera so that students can see the items cooking in the skillet. Turn on your heat source and keep the students back so they do not get burned.

Melting Butter:

Take the butter out of its packaging and show it to the students. Ask the students...

1) How does the butter appear?

2) How does the butter feel?

Place the butter in the hot skillet and have the students watch what happens. Ask them...

3) What do you see happening to the butter?

4) How does it appear?

When it is all melted, ask...

5) Do you think we can make the butter solid again? If so, how?

Carefully use a **hot pad**, **gloves**, or **hot hands** to hold the skillet's handle and pour the hot butter into a container that will hold it. Then place that container into a refrigerator. Later pull it out and show it to the students later after it has cooled.

Cooking an Egg:

Crack an egg, empty it into a clear container, and show the students. Ask them...

1) How does the egg appear?

Place the egg under the camera and have the students draw the raw egg. Then pour the egg into the skillet and have the students watch it cook. Ask them...

2) What do you see happening to the egg?

3) How does it look after it is cooked?

Have the students draw the cooked egg on some **paper**. Then ask them...

4) Can we make the egg raw again? Explain why.

Once something has been cooked, a chemical reaction has taken place, and it is now a new substance. Things taste different after being cooked from when they were raw. Many things are unsafe to eat when they are raw but safe to eat after they are cooked.

Cooking Popcorn:

Get some raw popcorn and show it to the students. Give them each some and ask...

1) How does it look?

2) How does it feel?

Have the students draw the raw popcorn on a piece of paper. Then if you still have some butter on your skillet, you can pour some popcorn into the skillet. And cook the popcorn. Make sure to have the camera filming the popcorn. Some should pop out and ask the students...

3) What did you see?

4) What force caused it to fly out of the skillet?

5) Do you think this was a push or a pull? Explain why.

6) How did this push cause the popcorn to move?

Explanation: When popcorn is popped, the liquid water inside the kernel is changed to steam. Pressure from the steam builds up inside the kernel. The kernel pops when the pressure gets too high, turning itself inside out (the soft starch on the inside becomes inflated and expands). The released pressure from steam forces the popcorn to move, causing it to fly. Ask the students…

7) Do you think we can make the popcorn go back to being a raw kernel? Explain why.

8) Give examples of changes caused by heat that can be reversed?

9) Give examples of changes caused by heat that cannot be reversed?

Dry it up

Directions:

Contest: Divide the class up into groups and have the students design a system to evaporate water at the fastest rate. They can use anything they could bring from home or that the teacher has in class. Each group will put an equal amount of water in a beaker and design a way to evaporate it as fast as possible.

Changing the Direction of Motion

Directions:

You will need some **safety goggles** and simple toy **Lego cars** for the students to follow the directions and put together. After their cars are assembled, ask them...

1) Which is more organized, the pile of Lego bricks or the car you made?

2) Which will be able to roll across the table, a pile of bricks, or the completed car?

3) Does your car move when nothing touches it?

4) How can you get your car to move?

5) How do you control the direction the car moves?

6) Can you make the car move in a direction it is not forced to go?

7) How can it change directions?

8) What happens if your Lego car crashes? Does that change its direction?

Have them put on their safety goggles and crash their cars, so they fall apart. Either roll them into a wall or roll them off the table, allowing them to crash on the floor. When objects move, they have Kinetic Energy (energy of motion); this energy can be transferred to cause the brick to come apart. Ask the students...

9) When a car crashes and breaks, does it work as it did before? Why.

10) Where did the kinetic energy of the car come from?

11) Can it be put back together to be able to roll again?

Have the students put their cars back together and roll them again. Then have the students take their cars completely apart and put them back into their containers.

Weather Patterns for the Seasons

Directions:

Draw and describe the four seasons showing the changes in nature in your region of the world. Make sure to describe how the plants look, temperature changes, and the length of days and nights. Then draw arrows showing the order the seasons occur.

Winter	
Fall	**Spring**
Summer	

Types of Dirt

Directions:

You need a **magnifying glass** for students to look at **sand**, **topsoil**, and **clay** samples. Have them draw what they see below.

Sand	Topsoil	Clay

1) What is the shape of each of the particles?

   a. Sand

   b. Topsoil

   c. Clay

2) Which particle has the biggest size?

3) Which particle has the smallest size?

4) How does each feel when you rub them between your fingers?

   a. Sand

   b. Topsoil

   c. Clay

Determining Soil Type

Directions:

You will need a **tablespoon**, a **soil sample**, a **beaker** with **water**, and a **pipette**. Place a tablespoon of soil in the palm of your hand. Add water a drop at a time and knead the soil to break down all aggregates. Soil has proper consistency when moldable, like putty. Use the flow diagram on the next page to find the type of soil you have by texture-by-feel- analysis. Answer the following questions.

1) Where did you collect the soil?

2) Did the soil remain in a ball when squeezed?

3) Did it form a ribbon?

4) How long was the ribbon?

5) Did it feel more gritty, smooth, or neither?

6) What type of soil do you have?

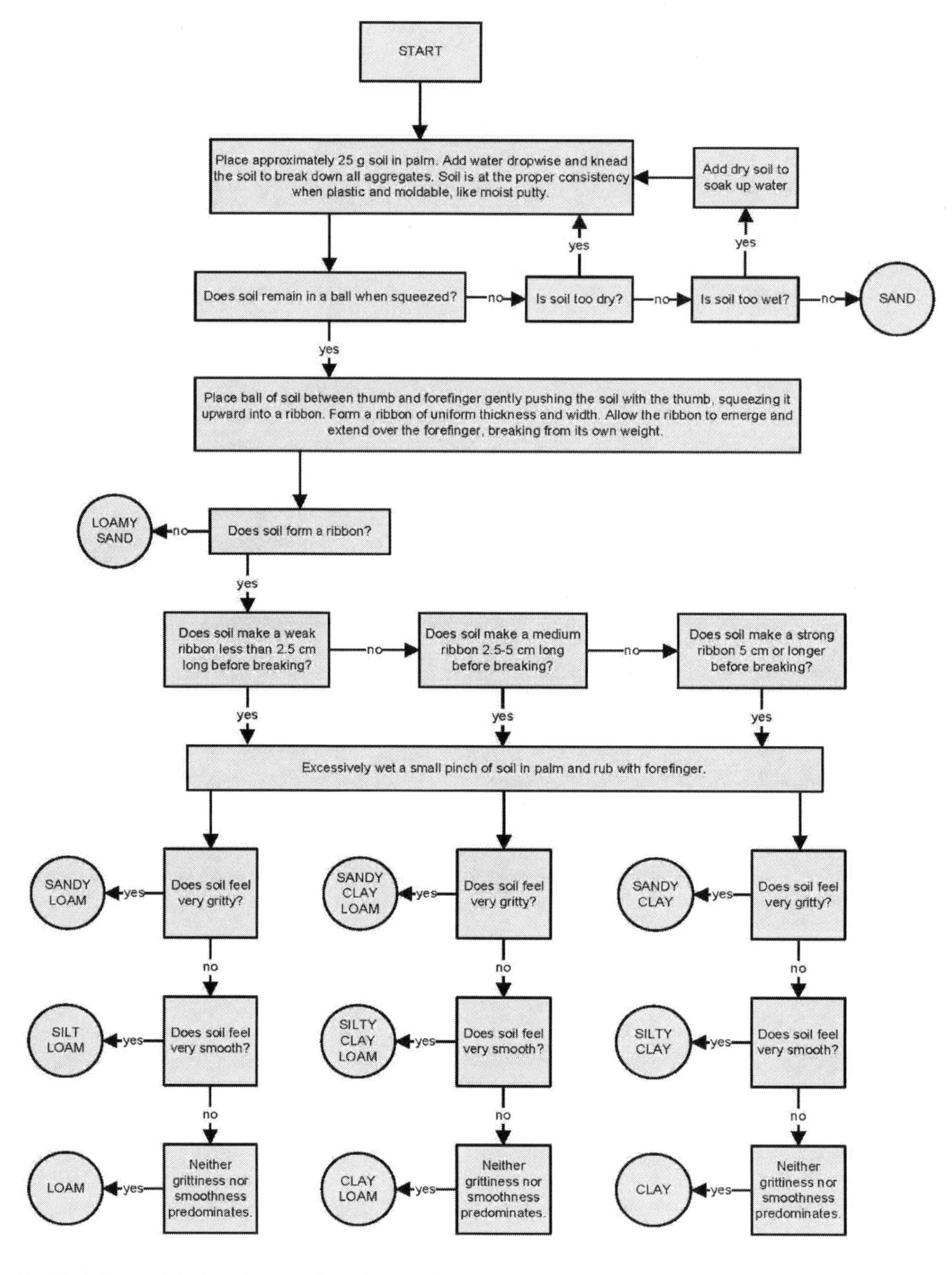

Modified from S.J. Thien. 1979. A flow diagram for teaching texture by feel analysis. Journal of Agronomic Education. 8:54-55.

Weathering

Directions:

You will need **safety goggles**, **sugar cubes,** and **chalk** to represent rocks, **water** in a **beaker**, **vinegar** in another **beaker**, and **pipettes**. Weathering is the breakdown of rock by wind, water, and chemical reactions.

1) The wind has particles that, when they hit a rock, microscopic pieces break off, making the rock smaller. Rub your hand across a sugar cube; what does this do to the cube?

2) Take another sugar cube and squirt water on it from a pipette. What does this do to the sugar cube?

3) What happens to water when it freezes?

   a. How can this action help break apart rocks?

4) Put on your safety goggles, take some chalk (representing limestone), and squirt water on it. Now take some vinegar (representing an acid) and squirt that on the chalk. What do you see chemically taking place?

5) How does this investigation model weathering?

6) What is not accurate about this model?

Erosion

Directions:

You will need an **apron**, **safety goggles**, **shovel**, **dirt** or **sand** in a **large tub/pan**, a **pitcher** of **water**, and a **spray bottle** that sprays water.

1) Go outside, dig up some dirt/sand, and put it into your large tub/pan. Fill it half full. Bring it inside to your lab table. Have some books raise up one side of the pan.
2) Take your spray bottle to simulate what a light rain might do to the dirt. Set it to mist and spray the dirt. This model simulates **sheet erosion**. How does the dirt move? Describe the pattern it is making.

3) Adjust your spray bottle to make a stream come out. This model will simulate heavier rain. Aim your bottle at an area of the dirt standing behind the high end of the tray and spray. How is the dirt movement different?

    a. What do you see forming in the dirt?

    b. Have multiple spray bottles spray at the same time. This model will simulate **rill erosion**. How is this pattern different than sheet erosion?

4) Now you will take your pitcher of water to simulate **gully erosion** and pour it from the high end of the tub. How is this dirt movement different from the others?

5) How did this investigation model erosion (the movement of particles)?

6) Take your wet dirt/sand back outside and dump it where your teacher tells you to. And dry out your tub/pan for the next class.

Bodies of Water

Directions:

Take the words below, cut them out, and have the students organize them into the groups mentioned below and have the students answer the questions that follow.

1) Make two piles; one pile is running water, and the other is water that stays in one place.
2) Then place them in order from smallest to largest.
3) Then make two new piles; one for freshwater and the other for saltwater.
4) Then make two new piles; one for long and skinny, and the other for wide.
5) Which bodies of water can be clear?

6) Which ones can be murky?

Take some **dirt** a place it in a **large beaker** of **water**. Stir it up. Then let it sit.

7) Can clarity change in a body of water? Explain how.

8) What are the different colors you have seen in each of these bodies of water?

9) What color is pure water?

10) Why do you think we see the different colors in different places and times?

Puddles	Ponds	Streams
Rivers	Lakes	Oceans

This page will be cut from the previous page.

Daily Weather Characteristics

Directions:

Ask the students...

1) How does weather affect our daily choices?

Have the class chart the weather characteristics for each day by placing a checkmark in the box if that day they experienced that characteristic.

Chart 1

Day	Hot	Cold	Clear	Cloudy	Calm	Windy	Rainy	Icy/snow
1								
2								
3								
4								
5								
6								
7								
8								
9								
10								
11								
12								
13								
14								
15								
16								
17								
18								
19								
20								

2) Which season are we in?

Repeat charting this investigation for different seasons. Then compare the different season charts with each other.

How Life uses Rocks, Soil, and Water

Directions:

Go outside and pick up a large rock. Then ask the students...

1) What do you see under it?

2) Was there more moisture above the rock or under the rock?

With a **shovel**, dig a small hole. Then ask the students...

3) What kind of animals did you see?

4) What part of the plants did you see?

5) Was there more moisture above the ground or below the ground?

6) What do you think some animals use rocks for?

7) What do you think animals use dirt for?

8) What do you think plants use dirt for?

9) What do plants use water for?

10) What do animals use water for?

Now have the students think about what they just observed and ask them...

11) What do humans use rocks for?

12) What do humans use dirt for?

13) What do humans use water for?

14) How do humans affect plants and animals?

Managing Garden Soil Moisture

Directions:

You will need a **heat lamp**, two shoebox-sized **tubs** of **soil** from the same source, one with some **mulch** covering the top and one without mulch. You will also need a **soil moisture probe** attached to an **interface** connected to a **computer** with **Logger Pro**.

1) Pour 500 mL of water evenly into each of the boxes of soil. Let it sit for a few minutes and measure the amount of soil moisture in each box by placing the soil moisture probe into the soil where the two prongs are vertical to each other and carefully pushing the probe into the soil until it is entirely under the soil. Write this data in Data Table 1.
2) Let the soil samples sit under a heat lamp for three-four days.
3) Then measure the moisture using the soil moisture probe again for each box of soil. Write this data in Data Table 1.

Data Table 1

	Soil With Mulch	Soil Without Mulch
Moisture Before		
Moisture After		
Change in Moisture		

Questions:

1) Which sample of soil held more moisture in it?

2) Which sample would you want to be set up for your bushes and flowers at your house? Explain Why.

3) What is the main purpose of mulch?

4) Which setup would allow you to use less water to keep your plants alive?

5) What are three materials in our area that are used for mulch?

6) Design an investigation to test the three types of mulch and see which one holds the most water in the same type of soil for a week. Use the soil moisture probe to see which mulch holds the most water in the soil. Which mulch held the most water?

Water Conservation

Directions:

Ask the students...

1) What do plants and animals use water for?

2) How do humans use water?

3) How do you see your parents use water?

Groundwater takes over 100 years to travel to its destination underground. Humans use lots of groundwater for agriculture and industry (they use the most water in the U.S. and the world). Surface water sources can be recharged as it rains. Only .1% of all water on Earth can be used by humans.

Fill a **beaker** with 1000 mL of **water** and put a drop of **blue food coloring** in it. Use a **pipette** and take out 1 mL of water. If the beaker represents all the water on the Earth, the pipette shows how much is available for humans to use.

4) What happens if it does not rain for a long time?

5) Why do we want to conserve water?

6) What are some ways we can conserve water?

Classifying Living and Nonliving Things

Directions:

Living things have basic needs to exist and produce young; nonliving things do not. Go outside and find ten things that are living and ten things that are not living and fill in the chart below.

Chart 1

Living Things	Nonliving Things
1	1
2	2
3	3
4	4
5	5
6	6
7	7
8	8
9	9
10	10

Competitive Relationships

Directions:

Biotic factors are living things; abiotic factors are nonliving things. Use the picture below to identify the model ecosystem's biotic and abiotic factors. Then discuss with your teacher and class the questions that follow and write down your answers.

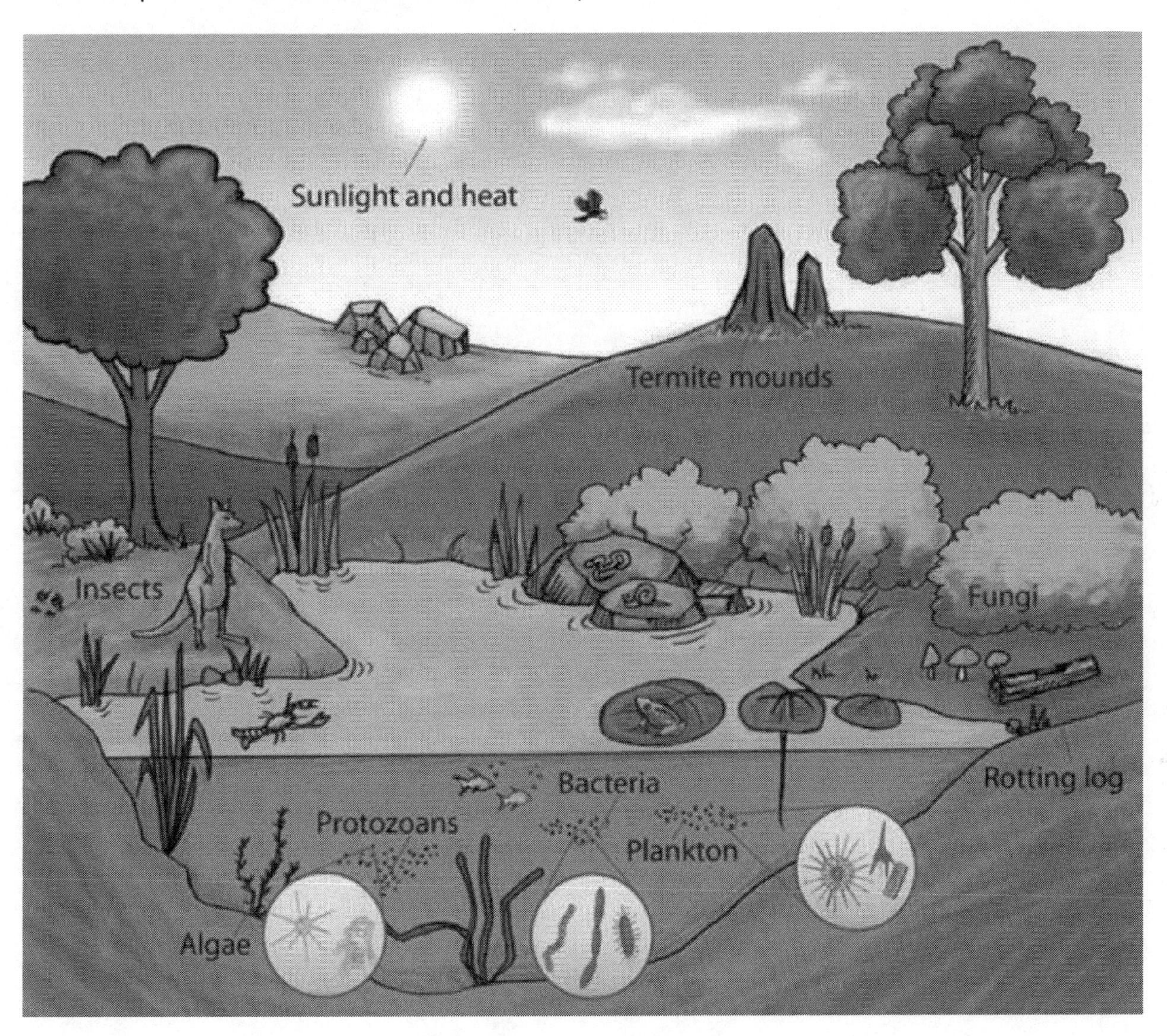

Picture from acamrmichael.weebly.com

Questions:

1) What are the biotic factors in this ecosystem?

2) What are the abiotic factors in this ecosystem?

3) What will plants compete for?

4) Which organisms here would be competing for soil?

5) Where might you find plants competing for light?

6) Which organisms are competing for water?

7) How do animals compete for food?

Who Eats Who?

Directions:

The arrows in a food web point to the organism that does the eating; use the food web below to see who is eating who in this ecosystem.

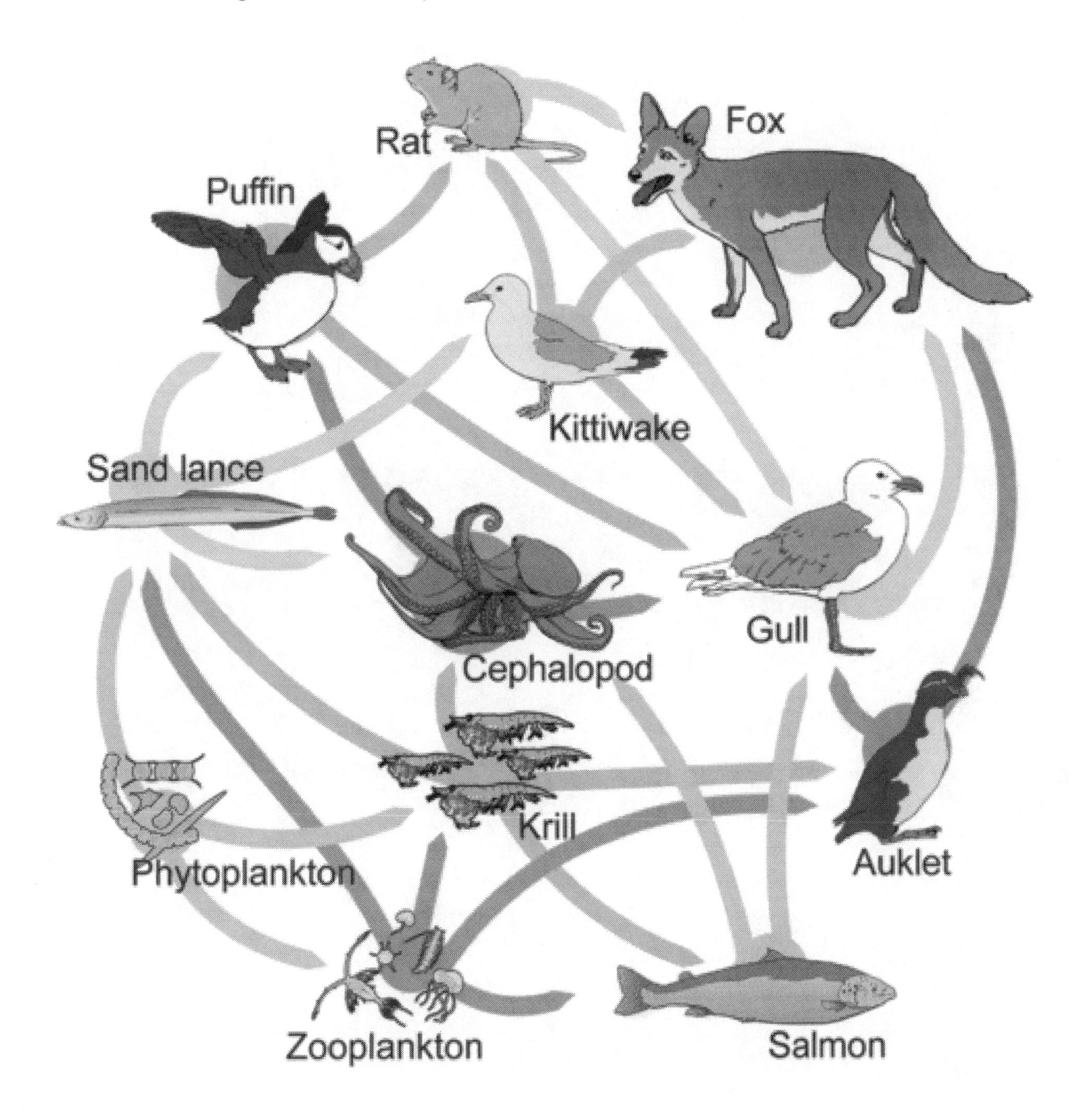

Picture from http://commons.wikimedia.org

1) The Phytoplankton use light, air, and water to make food; who eats the phytoplankton?
2) What do krill eat?
3) Who eats the krill?
4) What does the salmon eat?
5) Who eats the salmon?
6) What do puffins eat?
7) Who eats the puffin?
8) Which organism does not get eaten?
9) What do gulls eat?
10) Who eats the gull?
11) What do you think will happen to the foxes if very few phytoplankton are produced?
12) Show a food chain from the pictured web (one path from beginning to end).

Interactions of Living and Nonliving Things (Terrariums/Aquariums)

Directions:

Set up or see an **aquarium** or **terrarium** with **animal(s)** in it. Have the students observe and answer the questions below.

1) What living things do you see?

2) What are the nonliving things you see?

3) How do the nonliving things help the living things?

4) Do the living thing(s) help the nonliving things?

Bird Observations

Directions:

Watch birds and answer the questions.

1) What do birds use to move?

2) Which body parts help birds get food?

3) What body parts allow birds to detect predators and prey?

4) How do birds stay away from predators?

Squirrel Observations

Directions:

Watch squirrels and answer the questions.

1) What do squirrels use to move?

2) Which body parts help squirrels get food?

3) Which body parts help squirrels detect predators?

4) How do squirrels stay away from predators?

Fish Observations

Directions:

Watch fish in a pond or a fish tank and answer the questions.

1) What do fish use to move?

2) Which body parts help fish get food?

3) Which body parts allow fish to detect predators and prey?

4) How do fish keep from getting eaten?

Life Cycle of Chickens

Directions:

Draw a picture of an egg, a chick, and an adult hen to complete the life cycle of a chicken. Then answer the questions below.

Chick

Hen

Egg

1) Describe the life cycle of chickens.

2) How are baby chicks and young chickens different from adult chickens?

3) How does the baby chicken resemble its parent?

Squirrel Development

Directions:

Look at the picture to help you answer the questions below.

Picture from WordPress.com

1) How does the baby squirrel look like its parent?

2) How does the baby squirrel look different from its parent?

3) How will the baby squirrel change to become an adult?

Fish Life Cycle

Directions:

Label each part of the life cycle as adult fish, eggs, embryos, larvae, fry, and fingerlings. Then answer the questions that follow.

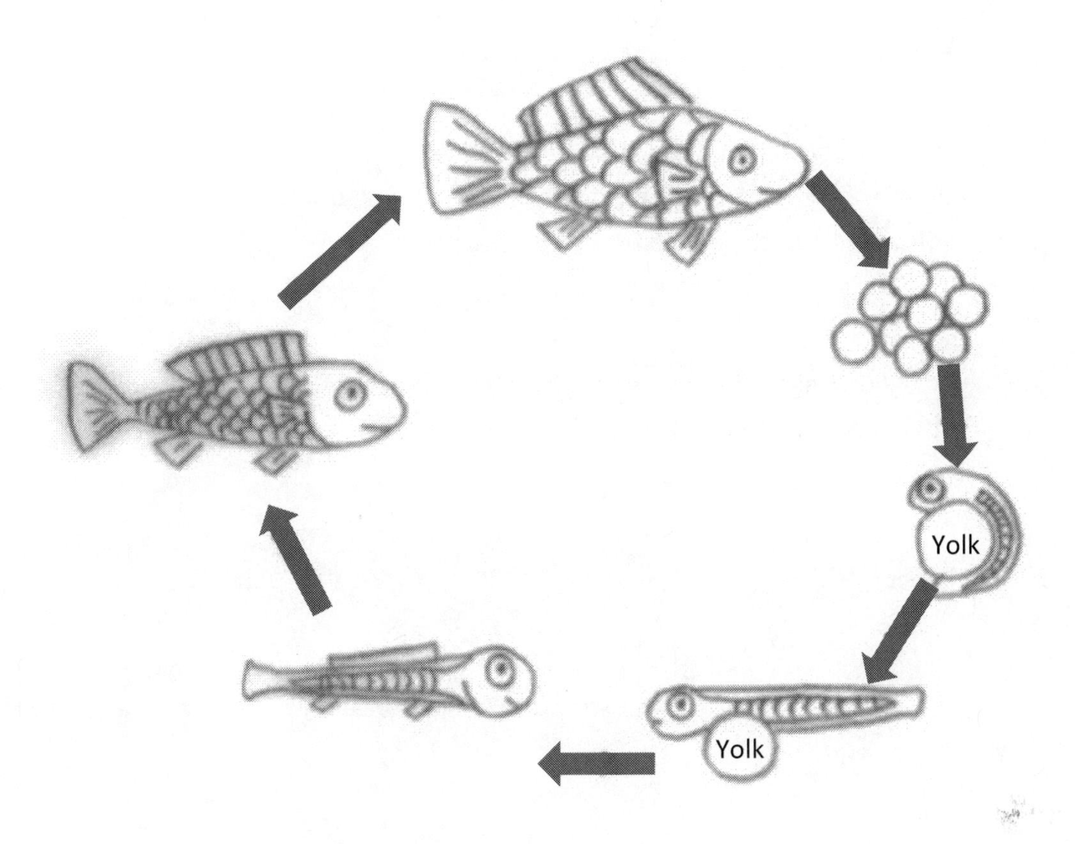

1) How do the larvae resemble their parents?

2) How do the larvae differ from their parents?

3) How are the fry different from the larvae?

4) How are the fingerlings different from the fry?

Sound Vibrates

Directions:

1) Take a **balloon** and blow it up. Talk next to it and feel the balloon vibrate. The balloon vibrates because air molecules are hitting it, showing the air vibrations. The balloon vibrates just like your eardrum vibrates.
    a. Talk into the balloon; what do you feel?

2) Tie two pieces of **string** on either side of a **wire hanger**.
    a. Wrap the string around each of your index fingers and clank the hanger against your desk. How does the hanger sound?

    b. Put your fingers in your ears and clank the wire hanger against the desk. How does the hanger sound now?

    c. Do sounds travel better through the air or the string?

    d. What if there is nothing for sound to travel through, will there be any sound?

    e. When the Death Star blows up in Star Wars, can that really make a sound in outer space? Why?

3) Take a **plastic tub** and three different **rubber bands** of the same length with different widths and wrap the rubber bands around the tub. Try to have the same tension on each rubber band.
    a. Pluck each rubber band and explain what you hear.

    b. Why do you think each rubber band makes a different sound?

Coffee Can Phones

Directions and Questions:

Take two **coffee cans** and poke a small hole in the bottom of each of them. Cut a long piece of **string** that reaches across the room, put the ends through each can, and tie a knot in them, fixing them to both cans.

1) Pull the string tight while holding the cans and talk through them. Can you be heard in the other can? Why do you think that is?

2) Have someone pinch the string with their fingers halfway across. Can you be heard in the other can now? Why do you think that is?

3) Let the string loosen and droop. Talk again. Can you be heard in the other can? Why do you think that is?

4) Combine with a group next to you, cross your strings, and have them touch while the strings are taught. Have someone talk into a can. Who can hear in their cans?

5) Have someone pinch with their fingers where the strings cross. Can anyone hear now?

6) Explain how sound travels from one can to the other.

Baby Signals

Directions:

Use **books** or the **internet** to have the students research baby behaviors that cause a response from the parents (Examples: offspring crying or chirping to have the parents feed or protect the offspring). Then answer these questions.

1) What was the animal you chose?

2) What did the baby do?

3) How did the parent respond?

4) How do the babies resemble their parents?

Then have the students share with the class what they found.

Building a Communication Device

Directions:

Students must design and build a nonelectronic device to communicate sound over a distance. Let students use materials from home or the teacher has available in the classroom. They can develop signals like Morris code or speak. Have the students then share how their devices communicate a message over a long distance to the class. Test them by seeing who can successfully send the most information over the longest distance. Information will be defined as the most letters in the alphabet used in the message.

Mimic Plants and Animals

Directions:

Human problems can be solved by mimicking a plant or animal. Have students use materials from around the house/classroom to solve a problem.

1) What was the problem?

2) What organism did you mimic?

3) How did you solve the problem?

Sun and Moon Daily Movement

Directions:

Use a **compass** to learn which direction is north, south, east, and west. Watch where the sun rises in the sky in the morning, then watch where it sets at night.

Sunrise & Sunset

1) Which direction did you see the sunrise?
2) Where do you think the sun will rise tomorrow?
3) Check and see. Does it come up in the same place every day?
4) Where would you expect the sun to be at noon each day?
5) Check and see. Is it there each day at noon?
6) Which direction did you see the sunset?
7) Where do you think the sun will set tomorrow?
8) Check and see. Does it set in the same place each evening?

Moonrise & Moonset

1) Which direction did you see the moon rise?
2) Where do you think the moon will rise tomorrow?
3) Check and see. Does it always rise in the same direction?
4) Where would you expect the moon to be at bedtime?
5) Check to see if it is there and stays there each night. Can you see the moon each night?
6) Which direction did you see the moon set?

7) Where do you think the moon will set tomorrow?

8) Check and see. Does the moon set in the same place?

Comparing the sun and moon

1) How are the moonrise and moonset different from the sunrise and sunset?

2) Can the sun be in the sky at night? Why.

3) Can the moon be in the sky during the day?

4) What defines day and night, the sun or the moon? Explain why.

Night and Day

Directions:

Ask the students:

1) What do you see in the sky during the day?
2) What do you see in the sky at night?
3) Have you ever seen the moon during the day?
4) Do you see the moon every night?
5) Why do you think you see the moon more at night than during the day?
6) What object provides light for us to see during the day?
7) What objects provide light for us to see at night?
8) Is the moon making light or reflecting light?
9) Where is the moon reflecting light from?
10) When is it warmer (day or night)?
11) When is it colder (day or night)?
12) Why is the day warmer than the night?
13) Does the sunset the same time every night?

Patterns with Sunrise and Sunset

Directions:

Have students find out when the sun rises and sets each day and what the high and low temperatures are. Chart this together as a class. You can use an almanac to find past data if you want. This chart will be useful in other lessons. A sample of a chart is below.

Date	Time Sunrise	Time Sunset	High Temperature	Low Temperature

Famous People in Science

Directions:

After students have performed some science investigations, have students watch videos or read to them stories about these famous people in science that have to do with what the students were investigating. Have them draw a picture of the person and describe what they are famous for.

Physics/Astronomy

Isaac Newton

Galileo Galilei

Nicolaus Copernicus

Johannes Kepler

Katherine Johnson

Astronauts

Buzz Aldrin

Neil Armstrong

Mae Jemison

Sally Ride

John Glen Allen Shepard

Biologists

Charles Darwin

Rachel Carson

Carl Linnaeus

Agnes Arber

Ynes Mexia

George Washington Carver

Gregor Mendel

Ernest Just

Alfred Russel Wallace

Antonie van Leeuwenhoek

I See Light Energy

Directions:

Ask the students the questions below and follow the instructions:

1) Do you think you can see your hand if no light shines on it?

2) The lights are on; can you see your hand now?

3) What color is it?

Turn off the lights.

4) Can you see your hand?

5) Does it have the same color it had when the lights were on?

If you are not in an interior room that blocks out most if not all light, find a place to take your students where they cannot see their hands in front of their face. Turn off the light for a couple of seconds, but not too long, so they do not get scared. After turning off the lights when they are looking at their hand and turning them back on, ask them...

6) Could you see your hand?

7) Why do you think you could not see your hand?

8) Did your hand have the same color as when the lights were on?

9) What do you think allows you to see objects?

10) What do you think allows you to see colors?

Tell the students: Light allows energy to move through space.

11) What do you think that energy can do?

If you have a **radiometer** (which can be found in Hobby Lobby), put it up in the window and have the students watch it when it is sunny and cloudy. Then sometime later, ask...

12) Why do you think the radiometer changes how it spins?

13) Which side, the black or white, is the light absorbed, and which side is it bouncing off?

    a. Black absorbs, adding energy, causing it to spin. The white cause the light to bounce off, reflecting energy away.

Traveling Light

Directions:

Have students look through the window. Ask them...

1) What do you see?

2) How is this different than looking through a wall?

3) Which object bocks light (a window or a wall)?

4) Which object allow light to travel through (a window or a wall)?

Turn on a **lamp** in the room with (no shade) and turn off all the other lights. Ask the students...

5) Where is your shadow?

6) Why is your shadow there?

7) Is your body allowing the light to pass through it or blocking it?

Have students hold some objects up to the light and see if they cast a shadow. Shadows will tell us if the object will block light or allow it to pass through. The darker the shadow, the more the object blocks light; the lighter the shadow, the less the object blocks light. Fill in Data Table 1 below and find objects that will let most light pass through, block some light, and block all light.

Data Table 1

Allow most light to pass	Block some light & allow some light to pass	Block all light

Virtual Investigations that go with 1st Grade Science

Explorelearning.com

Measuring Volume

Weight and Mass

Phases of Water

Energy Conversions

Weathering

Erosion Rates

Seasons in 3D

Water Cycle

Prairie Ecosystem

PhET.colorado.edu

Energy Forms and Changes

State of Matter: Basics

Force and Motion: Basics

Balancing Act

Waves Intro

Physicsclassroom.com/Physics-Interactives

Force

Kinetic Energy

Sound Wave Simulator

1st Grade TEKS and NGSS Correlations

Classifying Objects Science, Grade 1 (b) 1ABEF 5AC 6A

Ice to Water and Back Again Science, Grade 1 (b) 1ABCDE 3ABC 5ABE 6AB 8AB

Changing the Rate of Evaporation Science, Grade 1 (b) 1ABCDEF 2BC 3ABC 5ABCE 6AB 8A

Is Cooking Reversible? Science, Grade 1 (b) 1ABCDE 3ABC 5ABE 6AB 8AB

Dry it up Science, Grade 1 (b) 1ABCDEFG 2ABCD 4A 5ABCDEFG 6AB; K-2-ETS1-123

Changing the Direction of Motion Science, Grade 1 (b) 1ABCDE 3ABC 5ABDEFG 6C 7AB

Weather Patterns for the Seasons Science, Grade 1 (b) 1ABEFG 3AB 5AEG 9; 1-ESS1-2

Types of Dirt Science, Grade 1 (b) 1ABCDEFG 2B 3ABC 5AC 6A 10A

Determining Soil Type Science, Grade 1 (b) 1ABCDE 3AB 5AC 6A 10A

Weathering Science, Grade 1 (b) 1ABCDEG 2AD 3ABC 5ABCDG 6AB 10B

Erosion Science, Grade 1 (b) 1ABCDEG 2AD 3ABC 5ABCDEG 6A 10B

Bodies of Water Science, Grade 1 (b) 1ABG 2AB 3ABC 5AC 6A 10C

Daily Weather Characteristics Science, Grade 1 (b) 1ABEF 2BC 3B 5ACDEG 10D

How Life uses Rocks, Soil, and Water Science, Grade 1 (b) 1ABCDE 3ABC 5AFG 11A

Managing Garden Soil Moisture Science, Grade 1 (b) 1ABCDEFG 2ABC 3ABC 4A 5ABCEFG 6AB 11AC, K-2-ETS1-3

Water Conservation Science, Grade 1 (b) 1ABCDEG 2ABC 3ABC 5ACD 6A 11ABC

Classifying Living and Nonliving Things Science, Grade 1 (b) 1ABEF 2B 3B 5AD 12A

Competitive Relationships Science, Grade 1 (b) 1ABEG 3ABC 5BDEFG 11A 12AC

Who Eats Who? Science, Grade 1 (b) 1ABEG 3ABC 5BDEFG 12C

Interactions of Living and Nonliving Things Science, Grade 1 (b) 1ABCDE 3ABC 5DEFG 11A 12AB

Bird Observations Science, Grade 1 (b) 1ABCE 3ABC 5ABDF 7A 13AB

Squirrel Observations Science, Grade 1 (b) 1ABCE 3ABC 5ABDF 7A 13AB

Fish Observations Science, Grade 1 (b) 1ABCE 3ABC 5ABDF 7A 13AB

Life Cycle of Chickens Science, Grade 1 (b) 1ABEFG 2B 3AB 6A 5AG 13BC; 1-LS3-1

Squirrel Development Science, Grade 1 (b) 1ABE 2B 3ABC 5ACG 6A 13BC; 1-LS3-1

Fish Life Cycle Science, Grade 1 (b) 1ABEG 2B 3ABC 5AG 6A 13BC; 1-LS3-1

Sound Vibrates Science, Grade 1 (b) 1ABCDE 3ABC 5ABE; 1-PS4-1, K-2-ETS1-2

Coffee Can Phones Science, Grade 1 (b) 1ABCDE 3ABC 5ABE; 1-PS4-1, K-2-ETS1-2

Baby Signals Science, Grade 1 (b) 1ABE 3ABC 5BFG 13A; 1-LS1-2, 1-LS3-1

Building a Communication Device Science, Grade 1 (b) 1ABCDEG 2ABCD 5ACDEF; 1-PS4-4, K-2-ETS1-123

Mimic Plants and Animals Science, Grade 1 (b) 1ABCDG 2AD 3ABC 4A 5ABDF 6C 13A; 1-LS1-1, K-2-ETS1-12

Sun and Moon Daily Movement Science, Grade 1 (b) 1ABE 2B; 1-ESS1-1

Night and Day Science, Grade 1 (b) 1ABEF 2BC 3ABC 5ABCEFG; 1-ESS1-2

Famous People in Science Science, Grade 1 (b) 1AB 3AB 4AB

I See Light Energy Science, Grade 1 (b) 1ABCEG 2BC 3ABC 5ABCDEG; 1-PS4-2

Traveling Light Science, Grade 1 (b) 1ABCDEF 2BC 3ABC 4A 5ABCE; 1-PS4-3, K-2-ETS1-1

Equipment List for all 1st Grade Investigations

If you want to be able to do all the labs in this manual for 1st Grade Science, here is a list of all the equipment you will need (in order of appearance).

Ice cubes

Water

Beakers

Paper cups

Freezer

Glass/plastic containers

Heat lamp/incandescent bulb

Work lamp with an aluminum shield

Petri dishes

Video camera

Projector/smartboard

Hotplate/electric range

Skillet

Butter

Egg

Popcorn (raw)

Paper

Safety goggles

Lego car sets

Magnifying glasses

Sand

Topsoil

Clay

Tablespoons

Pipettes

Sugar cubes

Chalk

Vinegar

Aprons

Shovels

Large tub/pan

Pitcher

Spray bottle

Shoebox size tubs

Mulch

Soil moisture probe

Logger Pro/ digital program to analyze data

Food coloring

1000 mL beaker

Aquarium/terrarium

Balloons

String

Wire hanger

Coffee cans

Mirrors

Drum sticks

Flashlights

Horns

Bells

Compasses

Radiometer

lamp

2nd Grade Science Investigations

Slow Motion Collisions

Directions:

Take a video in slow motion mode with a **smartphone** of students hitting, kicking, or bouncing a **rubber ball**. Show the video and ask...

1) Before the collision, what was the motion of the ball?

2) During the collision, how did the shape of the ball change?

3) After the collision, how did the motion of the ball change?

4) What caused the ball to change its motion?

5) What effect does the change in shape have to do with the change in motion?

Bungee Rocket Launch

Directions:

Design an investigation using a **bungee rocket** to see how the rocket's motion will change with bigger and smaller pulls on the rocket. Use the table below to fill in your results.

Data Table 1

Analyzing Elastic Potential Energy

Directions:

You will need a **rubber band**, a **meter stick**, and a **foam disc**. Make marks on your table every 1 cm for 4 centimeters.

1) Place the rubber band on the zero mark so that the rubber band has no slack between two of your fingers.
2) Place your coin or disc in front of that rubber band, pull it back 1 cm, and release it.
3) Measure how far the disc traveled. Put this data in Data Table 1 below.
4) Repeat the procedures in #s 1-3, pulling the disc back 2 cm this time, and place this data in Data Table 1 below. Repeat this at 1 cm intervals, going longer until you pull it back 4 cm.

Data Table 1

Length Pulled Back (cm)	Distance Traveled (cm)
1 cm	
2 cm	
3 cm	
4 cm	

1) How did the disc's travel distance change as you pulled the disc farther back?

   a. Which pull-back distance made the disk go the farthest?

   b. Why do you think this happened?

Sound Vibrates

Directions:

1) Take a **balloon** and blow it up. Talk next to it and feel the balloon vibrate. The balloon vibrates because the air molecules are hitting it, showing air vibrations. The balloon vibrates just like your eardrum vibrates.
    a. Talk into the balloon; what do you feel?

2) Tie two pieces of **string** on either side of a **wire hanger**.
    a. Wrap the string around each of your index fingers and clank the hanger against your desk. How does the hanger sound?

    b. Put your fingers in your ears and clank the wire hanger against the desk. How does the hanger sound now?

    c. Do sounds travel better through the air or the string?

    d. What if there is nothing for sound to travel through, will there be any sound?

    e. When the Death Star blows up in Star Wars, can that really make a sound in outer space? Why?

3) Take a **plastic tub** and three different **rubber bands** of the same length with different widths and wrap the rubber bands around the tub. Try to have the same tension on each rubber band.
    a. Pluck each rubber band and tell what you hear.

    b. Why do you think each rubber band makes a different sound?

Coffee Can Phones

Directions:

Take two **coffee cans** and poke a small hole in the bottom of each of them. Cut a long piece of **string** that reaches across the room, put the ends through each can, and tie a knot in them, fixing them to both cans.

1) Pull the string tight while holding the cans and talk through them. Can you be heard in the other can? Why do you think that is?

2) Have someone pinch the string with their fingers halfway across. Can you be heard in the other can now? Why do you think that is?

3) Let the string loosen and droop. Talk again. Can you be heard in the other can? Why do you think that is?

4) Combine with a group next to you, cross your strings, and have them touch while the strings are taught. Have someone talk into a can. Who can hear in their cans?

5) Have someone pinch with their fingers where the strings cross. Can anyone hear now?

6) Explain how sound travels from one can to the other.

How is Sound Used?

Directions:

What are different types of sound that can be used in everyday life?

Type of Sound	Level of Sound	How it is used
	low/medium/high	
	low/medium/high	
	low/medium/high	
	low/medium/high	
	low/medium/high	
	low/medium/high	
	low/medium/high	
	low/medium/high	
	low/medium/high	
	low/medium/high	

1) What types of communication are soft?

2) What types of communication are loud?

Building a Communication Device

Directions:

Students must design and build a nonelectronic device to communicate sound over a distance. Let students use materials from home or the teacher has available in the classroom. They can develop signals like Morris code or speak. Have the students then share how their devices communicate a message over a long distance to the class. Test them by seeing who can successfully send the most information over the longest distance. Information will be defined as the most letters in the alphabet used in the message.

The Sun, Earth, and Moon

Directions:

The sun allows us to have days. The sun is the only star we can see during the day. It is the reason we cannot see other stars during the day. We will use a **lamp** to represent the sun, a **globe** to represent the Earth, and a **softball** to represent the moon. Turn off all the lights in the room and close the blinds so it is as dark as possible in the room. Turn on the lamp as the sole light source on one side of the room. Place the globe in the center of the room. Hold the softball outside where the students are sitting. We will use the lamp, globe, and softball to help answer the questions below.

1) Where is the dayside on Earth?

2) Where is the nightside of Earth?

3) If you spin the globe, how does that change day and night for life on Earth?

4) Where is the dayside on the moon?

5) Where is the nightside on the moon?

Move around the room with the softball and have the students tell you to stop when the whole moon would be visible from Earth, when half the moon would be visible, and when none of the moon would be visible from Earth.

6) How does the moon appear to change as it orbits the Earth?

7) Why does the moon change shape, cycling each month?

Telescopes and Binoculars

Directions:

Give the students some simple **telescopes/binoculars** to look through.

1) Have the students pick an object far from them in the room and draw what it looks like with their naked eyes.

2) Have them look through and focus the telescope/binocular on that same object and draw it again.

3) How does the second drawing differ from the first?

4) How do tools like telescopes and binoculars help us see objects far away?

5) Why do people use telescopes to look at our moon, other planets, and stars?

Major Weather Events

Directions:

Use the **internet** to fill in the chart while researching severe weather events.

Chart 1

Severe Weather	What it Does	When and where can they happen?
Hurricane		
Flood		
Tornado		

Weathering

Directions:

Weathering is the breakdown of rock by wind, water, and chemical reactions. Keep the objects inside the **pan/tray** as you do the investigation.

1) We will use **sugar cubes** to represent rocks. The wind has particles that, when they hit a rock, microscopic pieces break off, making the rock smaller. Our fingers will represent the wind. Rub your finger across a sugar cube; what does this do to the cube?

2) Take another sugar cube and squirt **water** on it with a **pipette**. What does this do to the sugar cube?

3) How can wind and water break apart rocks?

Erosion

Directions:

You will need an **apron**, **safety goggles**, **shovel**, **dirt**, or **sand** (sand will not make as much of a mess) in a **large tub/pan**, a **pitcher of water**, a **hairdryer**, and a **spray bottle** that sprays water.

Looking at the materials and lab we will be using, what safety precautions should we take to protect ourselves and materials during the investigation?

1) Go outside, dig up some dirt/sand, and put it into your large tub/pan. Fill it half full. Bring it inside to your lab table.
2) Angle your tub/pan so one end is higher than the other. We are going to simulate wind hitting the ground with a hairdryer. Put on your safety goggles. Plugin your hairdryer and keep it away from water. Turn on your hairdryer, have everyone in your group stand behind it so dirt does not blow on them, and bring the hairdryer closer to the dirt in your tub. What do you see happening to the small particles of dirt?

3) Unplug your hairdryer and set it aside. Take your spray bottle to simulate what a light rain might do to the dirt. Set it to mist and spray the dirt. How does the dirt move? Describe the pattern it is making.

4) Adjust your spray bottle to make a stream come out. This model will simulate heavier rain. Aim your bottle at an area of the dirt standing behind the high end of the tray and spray. How is the dirt movement different?

    a. What do you see forming in the dirt?

5) Now you will take a few rocks and place them in the middle of your tub/pan. Pour your pitcher of water from the high end toward the rocks to simulate a river. How is this dirt and rock movement different from the others?

Preventing Land Erosion

Directions:

Research the different ways we can slow or prevent the weathering and erosion of land. What can we use to hold the dirt together and shield it from wind and rain? Use a similar setup as in the last investigation (Erosion) and test your ideas.

1) What materials did you use?

2) How does your solution hold the ground together?

3) How does it block the wind and rain from directly hitting the dirt?

4) Test it; did it work?

    a. Are there any improvements you can make? If so, what are they?

Fast to Slow Land Change

Directions:

Have the students research the words below, cut them out, and place them in order from fastest to slowest to how the Earth's surface changes when these events happen. When the students are done, have them compare their answers and discuss why they put the words in the order they did.

Avalanche/mudslide	Earthquake	Erosion	Volcanic explosion

Modeling the Continental U.S.

Directions:

Use a map or globe of the Earth to see the shape of the United States. Use **clay** or **Play-Doh** to build a 3-D model of the continental United States, showing the shape of the land and the major bodies of water within its boundaries. Include mountain ranges, plains, major rivers, lakes, and where ice would be found.

Recording and Graphing Weather

Directions:

Use a **thermometer** to measure the outside temperature at the same time each day. Use a **rain gauge** to measure the amount of precipitation each day. Fill in Data Table 1 below for the Temperature and precipitation for each day. Then use the temperature data to fill in Graph 1 and the precipitation data to fill in Graph 2.

Data Table 1

Day	Temperature	Amount of Precipitation	Type of Precipitation
1			
2			
3			
4			
5			
6			
7			
8			
9			
10			

11			
12			
13			
14			
15			
16			
17			
18			
19			
20			

Graph 1

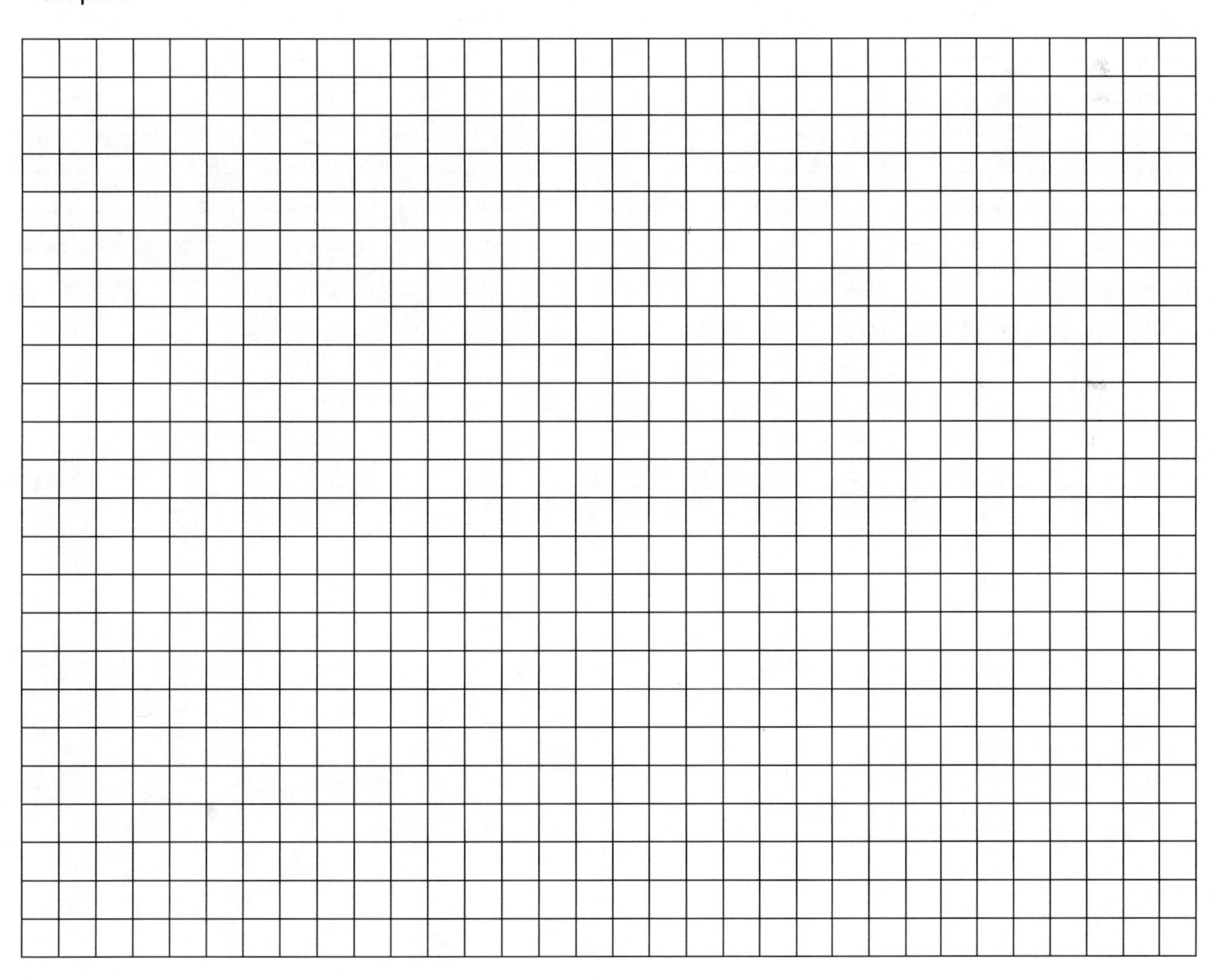

Graph 2

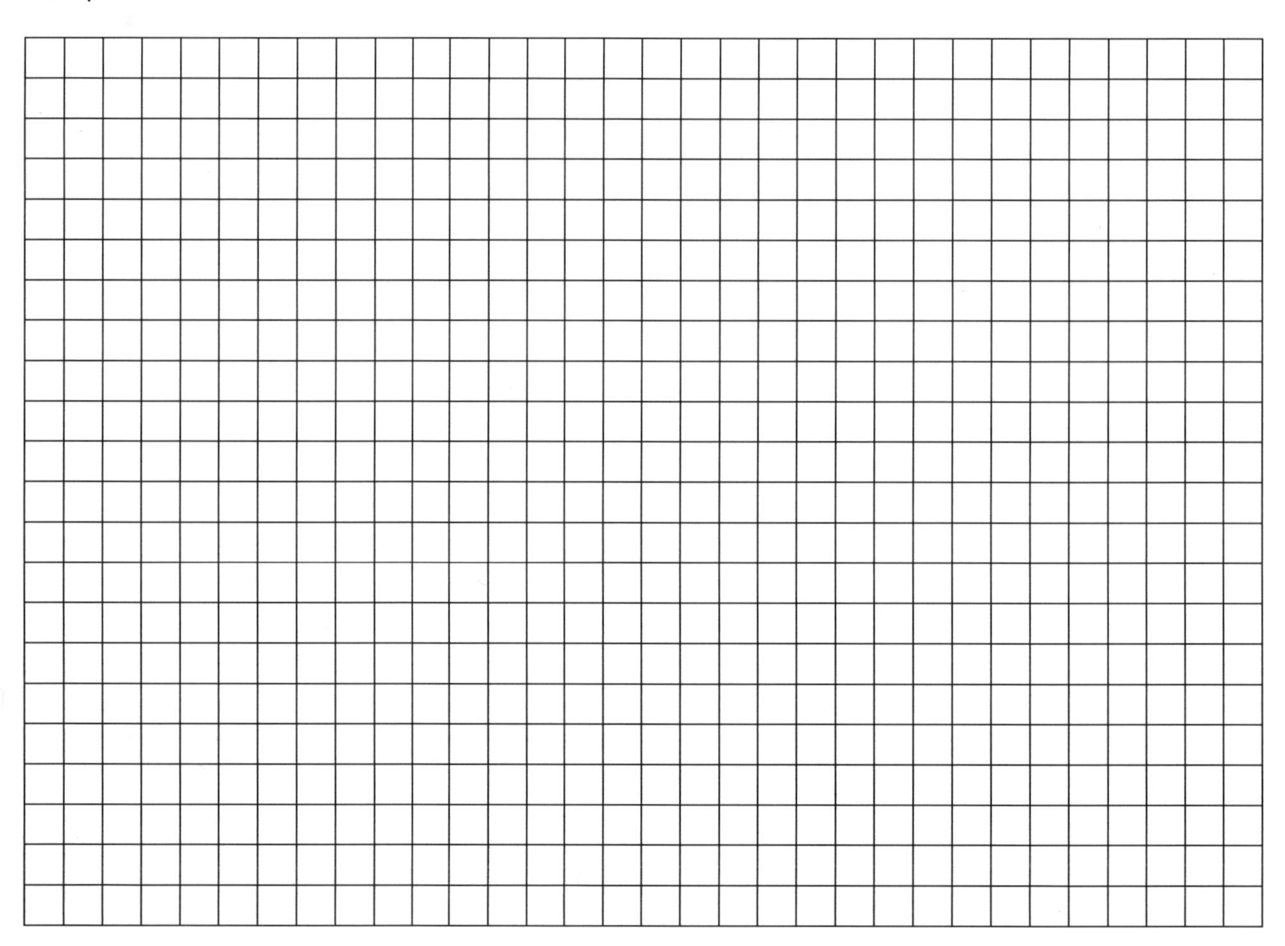

Questions:

1) How has the temperature changed from the first day to the last?
2) How is that affected by the season you were in?
3) Which day had the most rain?
4) How many days had no rain?
5) How many days had rain?
6) How could there have been an error in collecting the data?

Natural and Manmade Resources

Directions:

Natural resources come directly from nature to support life and meet peoples' needs. **Manmade resources** are when humans take those natural resources to produce something new that adds value to our lives. Use these definitions to cut out and divide the resources listed below into two piles **1) Natural Resources** and **2) Manmade Resources**.

Resources List

Agriculture	Fish	Plastic
Air	Forests	Roads
Air conditioners	Harbors	Rubber
Airplanes	Human population	Sand
Animals	Machinery	Soil
Birds	Metals	Stone
Bridges	Minerals	Sunlight
Cement	Money	Tools
Cities	Natural gas	Vehicles
Coal	Oil/petroleum	Watches
Electricity	Paper	Water
Factories	Phones	
Fertile land	Plants	

This page will be cut from the previous page.

Conserving Natural Resources

Directions:

Take your piles from the previous activity and think of ways to dispose of manmade products and natural resources to conserve natural resources. Examples could be replacing, reducing use, reusing, or recycling.

Completing Food Chains

Directions:

There are two food chains below. Above each organism, label it: producer, primary consumer, or secondary consumer. Then draw an arrow pointing the food to the animal that will eat it. Then below each organism, explain how they depend on other living things.

Biomes Chart

Use the **internet** to fill in the chart for the characteristics of each biome.

Biome	General Description	Support Plants	Support Animals	Amount of Rainfall
Tundra				
Temperate Coniferous Forest (Boreal Forest/Taiga)				
Temperate Deciduous Forest				
Grasslands				
Desert				

Biome	General Description	Support Plants	Support Animals	Amount of Rainfall
Chaparral				
Tropical Savana				
Tropical Rain Forest				

1) Of the different biomes above, which one has the most different kinds of life?

    a. Why do you think that is?

2) Which biomes have the least different kinds of life?

    a. Why do you think that is?

Plant's Dependence on an Ecosystem

Directions:

1) Plants pollinate to help produce the next generation of baby plants. They need a way to take the **pollen** (the male reproductive cell) to the **egg** (the female reproductive cell). Use the **internet** to research and find examples of how plants use wind, animals, and water to help pollinate plants. Fill in Chart 1 below to give two examples of each type of pollination.

Pollination Chart 1

	How use Wind	How use Animals	How use Water
Plant Example 1	Plant Name:	Plant Name:	Plant Name:
	Description:	Description:	Description:
Plant Example 2	Plant Name:	Plant Name:	Plant Name:
	Description:	Description:	Description:

Collect various fruits: **stickle burs**, **white puffed dandelions**, and **eatable fruit**. Have the students tell how each fruit is used to move the seeds away from the parent plant.

2) **Fruit** is defined as the part of the plant that helps move the seed away from the parent plant. Parent plants want their offspring far away from them, so they are not competing for the same resources. Some plants let the **wind blow away** the seeds, some **animals**

carry the seeds by sticking them to their fur or feathers, and others will have the **animals eat** the fruit to swallow the seeds and then **poop** the seeds out somewhere else. Fill in Chart 2 on the next page by drawing a picture of a fruit showing how it lets the wind blow it, takes a ride on an animal, or has an animal eat it to help move the seeds away from the parent.

Seed Distribution Chart 2

	Wind Blow Away	Animals Carry Away	Animals Eat & Poop
Picture of Fruit Example	Plant Name:	Plant Name:	Plant Name:

Modeling Animals Helping Plants Reproduce

Directions:

Contest 1: Students will build a model that mimics the function of an animal pollinating (vector pollinator) plants. **Plastic BBs** in a **container** (the teacher chooses) will mimic the pollen on the flower's anther (different flowers have different flower structures). The design that picks up the most individual pollen grains with a single touch and then drops them with a second movement wins.

Contest 2: Students will build a model that mimics the function of an animal dispersing seeds by contact with fur. **Plastic BBs** will also mimic the seeds. The students will build a fruit that contains seeds in it that sticks to fur as an animal walks by. The design that gets the most seeds picked up with a single swipe of fur across the fruit wins.

The models can be made of any materials the student can think up and bring from home or the teacher has in the classroom. Once the students have finished, have them present their models by showing them to the class how they help with either pollination or seed dispersal.

Plant Structures and Functions

Directions:

Find a picture of a plant where you see the roots, stems, leaves, flowers, fruit, and seeds. Print/cut out the picture, **paste** it below, and label its parts. Then next to each labeled plant part, explain how these structures help meet the plant's basic survival needs.

Do Plants Need Sunlight & Water?

Directions:

Plan and investigate whether **dry beans** planted in **cups** of **soil** need sunlight or **water** to grow.

1) Determine materials needed:

2) Form a Hypothesis:

3) Describe Experimental Group:

4) Describe Control Group:

5) Explain Results:

6) State Conclusion:

Animal Structures and Functions

Directions:

Pick an animal, find a picture, print it, or cut it out and **paste** it on this page. Label the animal structures and explain how they each are used to find and take in food, water, and oxygen.

Social Behavior

Directions:

Use the **internet** to research social behaviors and answer the following questions.

1) Why do birds flock, fish school, and cattle herd? What benefits do these behaviors give them?

    a. How does it help them defend themselves?

2) Why is it beneficial for animals of the same group to cooperate while hunting? Give examples.

3) Why is it beneficial for animals of the same group to cooperate while migrating? Give examples.

4) How does being in a group help animals cope with changes?

Butterfly Life Cycle

Directions:

Use the diagram below to answer the questions.

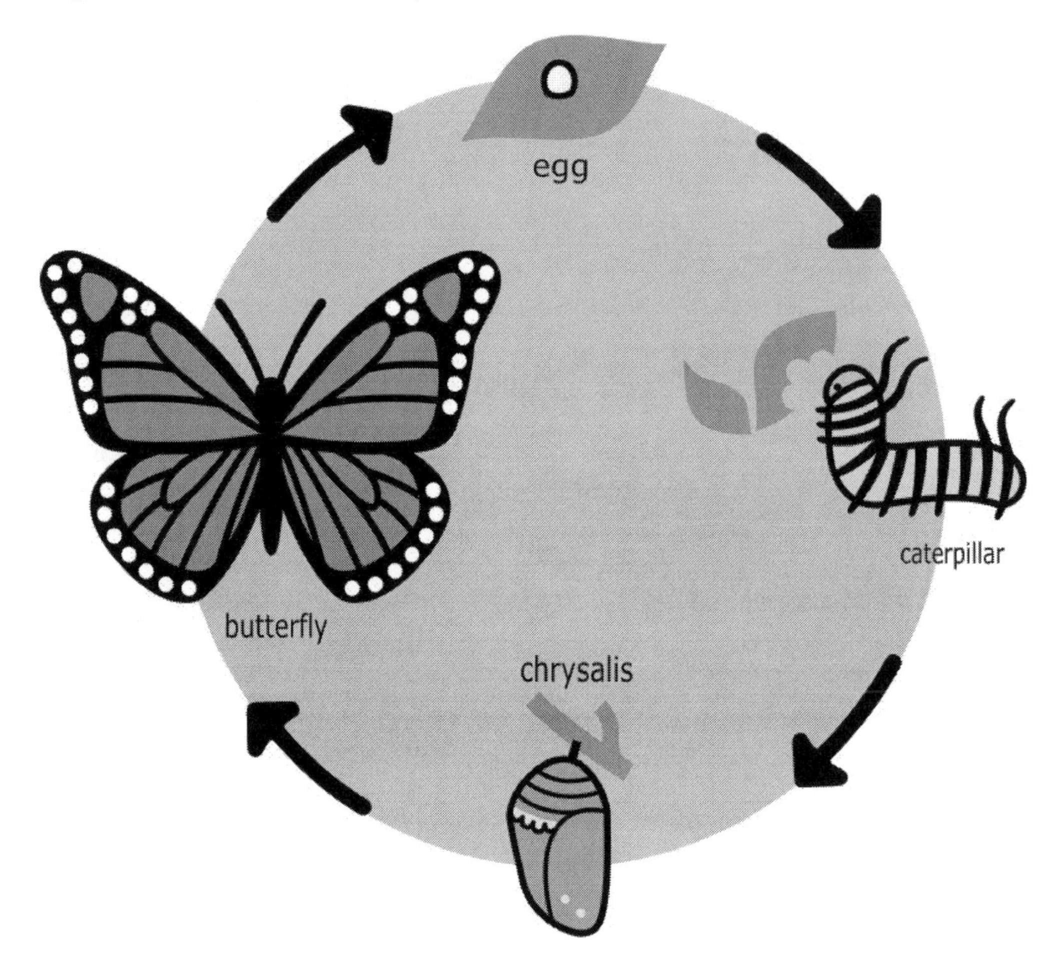

1) List the order in which the animal above develops from young to old.

2) How is the caterpillar different from a butterfly?

3) How is the egg different than the chrysalis?

4) What is happening in the chrysalis?

Frog Life Cycle

Directions:

Use the diagram below to answer the questions.

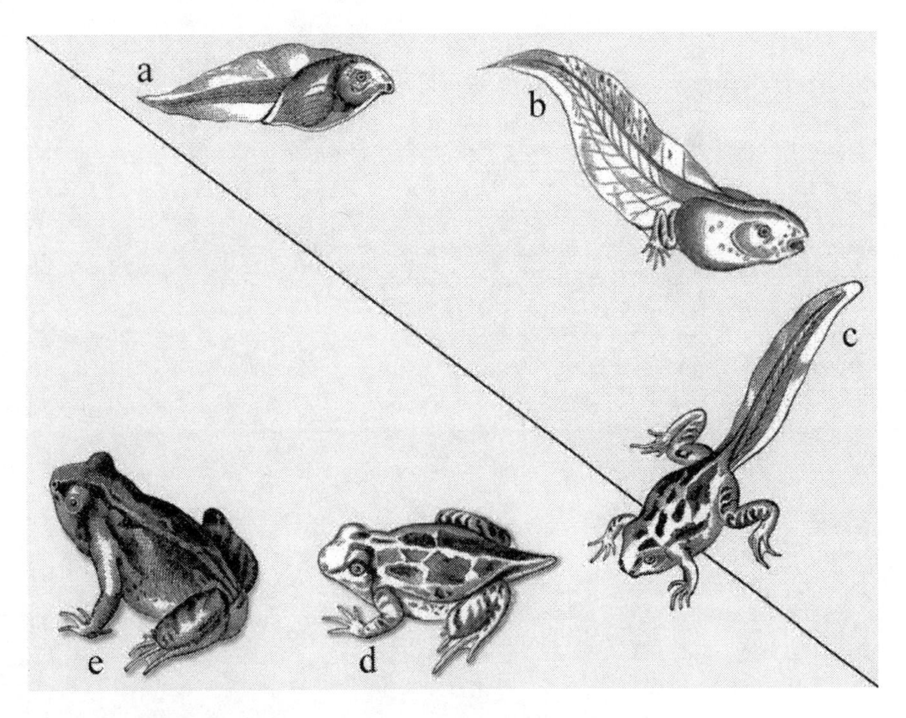

1) Which letter shows the animal above looks most like a fish?

2) Which letter is the animal an adult?

3) How does the animal in (a) look different from (e)?

4) Describe the changes that take place during a frog/toad's life cycle.

Lego Changes Challenge

Directions:

Take five **Legos** of the teacher's choice and see how many different shapes you can make using all five pieces. Draw the different shapes you made below. See which student can make the most shapes.

Classifying Objects

Directions:

Have students find objects that fit the physical properties below. Have them draw or write the objects that fit each property:

Red	Green	Rough	Smooth	Can cut with scissors	Easily folds

Changes when frozen	Changes when melts	Hard	Soft	Stiff	Flexible

1) Which qualities can wood material possess?

2) Which qualities can wood materials not possess?

3) How can wood change properties from rough to smooth and stiff to flexible?

Changing Properties

Directions:

Give each student two pieces of paper, scissors, and a clear cup/glass with an ice cube.

1) Cut a piece of paper into smaller pieces. Which properties were changed?

    a. Which properties were not changed?

2) Fold a piece of paper multiple times. Which properties were changed?

    a. Which properties were not changed?

3) Observe an ice cube melt. Which properties changed?

    a. Which properties did not change?

What Could I Use?

Directions:

Find different objects in the room or outside that could be used to make the tools needed below. List the objects you used to make each tool and describe how you put them together.

1) Fishing pole:

2) Splint for a broken finger:

3) Suspenders:

4) Fork/spoon for picking up food:

5) Broom and dustpan:

6) Checkers/chess pieces and board:

7) Envelope:

8) Pendulum:

9) Wheel and axel:

10) Teacher's choice:

Famous People in Science

Directions:

After students have performed some science investigations, have students watch videos or read stories about these famous people in science that have to do with what the students were investigating. Have them draw a picture of the person and describe what they are famous for.

<u>Inventors</u>

Alexander Graham Bell

Stephanie Kwolek

Thomas Edison

Marie Van Brittan Brown

George Washington Carver

Maria E. Beasley

Benjamin Franklin

<u>Biologists</u>

Charles Darwin

Rachel Carson

Carl Linnaeus

Agnes Arber

Ynes Mexia

Gregor Mendel

Ernest Just

Alfred Russel Wallace

Antonie van Leeuwenhoek

Jane Goodall

<u>Chemists</u>

Marie Daly

Mario Molina

Marie Curie

Dimitri Mendeleev

John Dalton

Rosalind Franklin

Ice to Water and Back Again

Directions:

You will need **ice cubes** floating in **water** inside a large clear **beaker** positioned where all students can see. Then ask the students these questions...

1) What do you think ice is?

2) How does ice feel?

3) Is ice lighter or heavier than liquid water? Explain how you know.

4) What do you think will happen if we let it sit for a while?

5) What is appearing on the outside of the container that we must wipe off to see inside?

6) Where do you think it is coming from?

Let it sit and melt. After it melts, bring the student's attention back to it and ask...

7) Where did the ice go?

8) What is ice? How do we know?

9) How does water feel?

10) Can we make water into ice? If so, how?

Have students put their names on **paper cups**. Pour the water into their cups and have them place them in a **freezer**. Come back at the end of the day or the next day and see what happened to their water. Then ask the students...

11) What happened to the water?

12) How do we know this is ice?

a. How does ice feel?

b. Does this float in liquid water?

13) Can we change it back to liquid water? If so, how?

14) What else do you think can happen to water?

Pour water into a beaker and make a mark on the container where the water's surface is. Let it sit out for a few days and have the students see the water level drop. Then ask the students...

15) What do you think happened to the water?

16) Where did it go?

Is Cooking Reversible?

Directions:

You will need a **video camera** connected to a **smartboard** or **some other form of projection** to safely show the action of cooking taking place. You will want a **hotplate** or an **electric range** and a **skillet** to melt **butter**, cook an **egg**, and cook **popcorn**. Line up the camera so that students can see the items cooking in the skillet. Turn on your heat source and keep the students back so they do not get burned.

Melting Butter:

Take the butter out of its packaging and show it to the students. Ask the students...

1) How does the butter look?

2) How does the butter feel?

Place the butter in the hot skillet and have the students watch what happens. Ask them...

3) What do you see happening to the butter?

4) How does it look?

When it is all melted, ask...

5) Do you think we can make the butter solid again? If so, how?

Carefully use a **hot pad**, **gloves**, or **hot hands** to hold the skillet's handle and pour the hot butter into a container that will hold it. Then place that container into a refrigerator. Later pull it out and show it to the students later after it has cooled.

Cooking an Egg:

Crack an egg, empty it into a clear container, and show the students. Ask them...

1) How does the egg appear?

Place the egg under the camera and have the students draw the raw egg below. Then pour the egg into the skillet and have the students watch it cook. Then have them draw the cooked egg below.

Raw Egg	Cooked Egg

2) What do you see happening to the egg?

3) How does it look after it is cooked?

4) Can we make the egg raw again? Explain why.

Once something has been cooked, a chemical reaction has taken place, and it is now a new substance. Things taste different after being cooked from when they were raw. Many things are unsafe to eat when they are raw but safe to eat after they are cooked.

Cooking Popcorn:

Get some raw popcorn and show it to the students. Give them each some and ask...

1) How does it look?

2) How does it feel?

Have the students draw the raw popcorn below. Then if you still have some butter on your skillet, you can pour some popcorn into the skillet. And cook the popcorn. Make sure to have the camera filming the popcorn. Some should pop out and have the students draw it below.

Raw Popcorn	Cooked Popcorn

3) What did you see?

4) What force caused it to fly out of the skillet?

5) Do you think this was a push or a pull? Explain why.

6) How did this push cause the popcorn to move?

Explanation: When popcorn is popped, the liquid water inside the kernel is changed to steam. Pressure from the steam builds up inside the kernel. The kernel pops when the pressure gets too high, turning itself inside out (the soft starch on the inside becomes inflated and expands). The released pressure from steam forces the popcorn to move, causing it to fly. Ask the students...

7) Do you think we can make the popcorn go back to being a raw kernel? Explain why.

8) Give examples of changes caused by heat that can be reversed?

9) Give examples of changes caused by heat that cannot be reversed?

Virtual Investigations that go with 2nd Grade Science

ExploreLearning.com

Measuring Volume

Weight and Mass

Growing Plants

Phases of Water

Energy Conversions

Phases of Moon

Hurricane Motion

Prairie Ecosystem

Honeybee Hive

Weathering

Erosion Rates

Food Chain

Animal Group Behavior

PhET.colorado.edu

Density

Energy Forms and Changes

States of Matter: Basics

Force and Motion: Basics

Balancing Act

Waves Intro

Physicsclassroom.com/Physics-Interactives

Force

Kinetic Energy

Simple Wave Simulator

2nd Grade TEKS and NGSS Correlations

Slow Motion Collisions Science, Grade 2 (b) 1ABCDE 3ABC 4A 5ABEG 7AB

Bungee Rocket Launch Science, Grade 2 (b) 1ABCDEF 2BC 5ABCE 7B

Analyzing Elastic Potential Energy Science, Grade 2 (b) 1ABCDEF 2BC 3AB 5ABCE 7B

Sound Vibrates Science, Grade 2 (b) 1ABCDE 2B 3ABC 5ABCE 8A; K-2-ETS1-2

Coffee Can Phones Science, Grade 2 (b) 1ABCDE 3AB 5ABDEF 8AC; K-2-ETS1-2

How is Sound Used? Science, Grade 2 (b) 1ABF 3AB 8B; K-2-ETS1-12

Building a Communication Device Science, Grade 2 (b) 1ABCDEG 2ABCD 5ACDEF 8AC; K-2-ETS1-123

The Sun, Earth, and Moon Science, Grade 2 (b) 1ABCDEG 2AB 3ABC 5ABDEF 9A; K-2-ETS1-2

Telescopes and Binoculars Science, Grade 2 (b) 1ABCDF 2B 3ABC 4A 5AF 9B

Major Weather Events Science, Grade 2 (b) 1ABEF 3AB 5ADE 10C

Weathering Science, Grade 2 (b) 1ABCDEG 2A 3AB 5ABDE 10A

Erosion Science, Grade 2 (b) 1ABCDEG 2A 3AB 5ABDE 10A; 2-ESS1-1

Preventing Land Erosion Science, Grade 2 (b) 1ABCDE 2AD3ABC 4A 5ABDEF 10A; 2-ESS2-1, K-2-ETS1-12

Fast to Slow Land Change Science, Grade 2 (b) 1ABF 2B 3ABC; 2-ESS1-1

Modeling the Continental U.S. Science, Grade 2 (b) 1ABCDEG 2ABCD 3ABC 5CD; 2-ESS2-23, K-2-ETS1-2

Recording and Graphing Weather Science, Grade 2 (b) 1ABCDEF 2BC 3AB 5ABC 6A 10B

Natural and Manmade Resources Science, Grade 2 (b) 1ABEF 5A 11A

Conserving Natural Resources Science, Grade 2 (b) 1AB 5A 11B

Completing Food Chains Science, Grade 2 (b) 1ABFG 2B 5ABDEF 12B

Biomes Charts Science, Grade 2 (b) 1ABEF 3AB 5ACDEF 12A; 2-LS4-1

Plant's Dependence on an Ecosystem Science, Grade 2 (b) 1ABEF 3AB 5ABDF 12C

Modeling Animals Helping Plants Reproduce Science, Grade 2 (b) 1ABCDG 2ABCD 3BC 4A 5AF 12C; 2-LS2-2, K-2-ETS1-123

Plant Structures and Functions 1 Science, Grade 2 (b) 1ABEFG 5DFG 13A

Do Plants Need Sunlight and Water? Science, Grade 2 (b) 1ABCDEF 2BCD 3ABC 5ABCDEFG 13A; 2-LS2-1

Animal Structures and Functions 1 Science, Grade 2 (b) 1ABEFG 5DFG 13B

Social Behavior Science, Grade 2 (b) 1ABE 3ABC 5ABFG 13C

Butterfly Life Cycle Science, Grade 2 (b) 1ABEG 2B 3ABC 5ADFG 13D

Frog Life Cycle Science, Grade 2 (b) 1ABEG 2B 3ABC 5ADFG 13D

Lego Changes Challenge Science, Grade 2 (b) 1ABCDEF 3AB 5ACDG 6C; 2-PS1-3

Classifying Objects Science, Grade 2 (b) 1ABEF 5C 6AB; 2-PS2-12

Changing Properties Science, Grade 2 (b) 1ABCDE 2BC 3ABC 5ABCG 6AB; 2-PS1-1

What Could I Use? Science, Grade 2 (b) 1ABCDEG 2AD 3AB 5ADF 6AC 7A; 2-PS1-2, K-2-ETS1-2

Famous People in Science Science, Grade 2 (b) 1AB 3AB 4AB

Ice to Water and Back Again Science, Grade 2 (b) 1ABCDE 3ABC 5ABE 6AB; 2-PS1-4

Is Cooking Reversible? Science, Grade 2 (b) 1ABCDE 3ABC 5ABE 6AB; 2-PS1-4

Equipment List for all 2nd Grade Investigations

If you want to be able to do all the labs in this manual for 2nd Grade Science, here is a list of all the equipment you will need (in order of appearance).

Smartphone

Rubber balls

Bungee rockets

Meter sticks

Foam discs

Balloons

String

Wire hangers

Plastic tubs

Coffee cans

Lamp

Globe

Softball

Telescopes/binoculars

Internet

Pans/trays

Sugar cubes

Water

Pipettes

Aprons

Safety goggles

Shovels

Dirt/soil

Large tubs/pans

Pitchers

Hairdryers

Spray bottles

Clay/Play-Doh

Thermometer

Rain gauge

Stickle burs

White puffed dandelions

Eatable fruit

Paste

Dry beans

Cups

Legos

Paper

Scissors

Hotplate/electric range

Skillet

Butter

Eggs

Popcorn

Water

Ice cubes

Beakers

Freezer

3rd Grade Science Investigations

What Causes Sinking and Floating?

Directions:

You will use a **scale**, a **ruler**, a **graduated cylinder**, and the outline below to design an investigation to determine why objects either sink or float in water.

1) **Problem:** What do you think causes objects to sink and float?
2) What do you think the answer is? Write a **Hypothesis:**

3) Describe what you will measure with the tools to solve the problem in the **Experimental Setup:**

4) Show your data here as **Results:**

5) What patterns do you see in the data that solved the problem in #1? Explain it in the **Conclusion:**

Build a Useful Boat

Directions:

Use **oil-based clay** and other objects in the room (which cannot exceed the mass of the clay) to build a boat to hold the heaviest load in a **bucket/tub** of **water** and not sink. Everyone has to start with the same amount of clay. The clay must touch the water. Put each boat in the water and put standard masses into the boat to find the winner. Draw your final design below and write down how much weight your boat held before it sank. Compare designs to see what the best performing boats had in common.

Amount of clay your teacher allowed: ____

Your design:

Maximum amount of mass held:____

Solids, Liquids, and Gases

Directions:

Look at a **plastic block(s)** in a **beaker**, **water** with **food coloring** in a **beaker**, and **dry ice** (use **gloves** when touching dry ice, never let it touch your skin) in an **Erlenmeyer flask** with a cork in it. If you need to, add some water to the dry ice to get the cloud to form faster. The block is solid, the colored water is a liquid, and the cloud coming off the dry ice is carbon dioxide gas. After looking at the contents in the beakers and flask, answer the following questions.

1) How is the shape of a solid different from liquid and gas?

2) How is the shape of a liquid different from a gas?

3) Which of the three phases has a definite shape?

4) Which of the three takes the shape of the container?

5) Which of the three will fill the whole container?

6) Besides the block(s) in the beaker, where else do you see solids in the investigation?

7) Where do you see a solid becoming gas?

8) How could you change the liquid into a solid?

9) How could you change the liquid into a gas?

10) Why do you think the dry ice changes from a solid straight to a gas without being a liquid?

Magic Rings

Directions:

Before any students are around: Take a **long golf tee** and **glue** a **ring magnet** to where the tee is going through the hole in the magnet, and the tee will stand up when flipped upsidedown. After the glue dries, place another **ring magnet** on top of the other magnet to have it stick to the bottom magnet.

With the students:

1) Take the two magnets with the golf tee sticking through them and show them what you have without saying anything.
2) Take the top magnet off and ask a student to throw an imaginary spring onto the golf tee. Act as if you caught it on the tee. Make sure you flipped the magnet you pulled off over so it will repel the bottom magnet. Place the top magnet on and show that it now bounces and hovers in the air on the tee (don't bounce it too much, so the magnet does not fall off and break on the floor).
3) Take the top magnet off, flip it back in your hand, and have a student pretend to take the imaginary spring off. Place the top magnet back on so that it will stick to the bottom magnet.
4) Ask the students how they think that happened. Repeat steps 1-3 again until they guess correctly. Then show them that they are ring magnets and can attract or repel each other depending on how they are flipped (facing each other).
   a. Explain that this is a **magnetic force** which is one of the major forces used in nature to make the world work.
   b. Ask: Is a magnetic force a push or a pull?

      i. How do you know?

5) How can the trains at Disney World hover above the rail it is riding on without touching the rail?

Condensation of Water

Directions:

Look at a **beaker** of **ice water**. Then answer the following questions.

1) What do you see forming on the outside of the glass?

    a. When the temperature drops, gases can change to a liquid. Where do you think it came from?

2) How is a liquid different from a gas?

3) Where have you seen a force pull objects together?

4) Blow up a **balloon** and tie it closed. Rub it on your hair, then stick it to your shirt. What do water drops on the glass and the balloon on your shirt have in common?

5) Place the charged balloon next to a small stream of water. How does the charged balloon affect the water stream?

6) What do you think allows the water to stick together on the glass?

Rolling Brick Tower

Directions:

Part 1: Have the students use **Mega Blocks** to build the highest tower on the rolling block (block with wheels) without falling over as you push it. Have the students experiment with different patterns to build the strongest tallest tower on wheels.

1) How high was the tallest tower in the class?

2) What pattern was used to make the tower stronger?

Part 2: Have students discuss how they could add **masking tape** to their towers to make them stronger. Add the tape to strengthen their towers and build them higher.

1) How tall was the tallest tower after they added tape?

2) How was the tape used to help make the towers taller?

3) Could other materials be added to make the towers stronger so they could be made even taller? If so, what are they?

4) How could this information be useful in making a building resistant to earthquakes?

Let's Swing

Directions:

Go out to the **swing set** on the playground and have students sit on the swing(s).

1) Describe the motion of the students just sitting on the swings.

    a. Are the forces acting on the person and swing balanced or unbalanced?

2) How could their motion change?

    a. Would they get a push or a pull?

3) While swinging on the swing set, how does the motion change?

    a. What part of the swing is the student speeding up?

    b. What part of the swing is the student slowing down?

    c. When did the student stop for a split second?

    d. When is the student moving the fastest?

    e. When is the student moving the slowest?

4) What force is causing the swing to accelerate down?

Let's Swing Again

Directions:

The faster an object moves, the more **kinetic energy** (energy of motion) it has. The higher an object gets, the more **potential energy** (ability to go into motion) it has. Go out to the **swing set** on the playground and have students sit on the swing(s).

The total mechanical energy = potential energy + kinetic energy (ME = PE + KE)

1) How much mechanical energy does the student have just sitting there?

2) Have the students swing. What part of the swing does the student have the most kinetic energy (energy of motion)?

3) Which part of the swing does the student have the least kinetic energy?

4) Which part of the swing has the most potential energy (ability to go into motion)?

5) Which part of the swing has the least potential energy?

6) What happens to the kinetic energy when the potential energy increases in the swing?

7) What happens to the kinetic energy when the potential energy decreases in the swing?

8) How is the motion of the swing predictable?

9) Does the swing in motion have balanced or unbalanced forces? Explain why.

Bouncing Ball

Directions:

Take a **rubber ball** and bounce it. **Kinetic Energy** is the energy of motion. **Potential Energy** is the ability to have motion.

1) When does the ball have the most kinetic energy?

2) When does the ball have the most potential energy?

3) When does the ball have the least kinetic energy?

4) When does the ball have the least potential energy?

5) What happens to the potential energy when the kinetic energy increases?

6) What happens to the potential energy when the kinetic energy decreases?

7) What do you hear when you bounce the ball?

8) Why do you think the bounces get smaller and smaller when you let the ball bounce by itself?

9) What is predictable about the motion of the bouncing ball?

10) Are the forces acting on the bouncing ball balanced or unbalanced? Explain why.

Types of Energy

Directions:

Observe these objects live or on the **internet** and tell what types of energy they have from the list below. After you tell what type of energy the object has, explain how you know it has that type of energy.

Light Energy Thermal Energy Mechanical Energy

1) A car being driven down the street
2) A student swinging
3) The sun
4) A star
5) Running water
6) Light bulbs turned on
7) A pencil rolling on the table
8) Blacktop on a sunny day
9) A person running

Map of our Solar System

Directions:

Use the paper below, the **internet**, and **colored pencils** to draw a map of how the planets and their moons orbit the sun in our solar system (It does not have to be to scale).

How the Earth and Moon Orbit the Sun

Directions:

Part 1: Have the students divide into groups of three. Have one represent the sun, another represents the Earth, and the last student represents the moon. Have the Sun sit still, have the Earth walk around the sun, and have the moon walk around the Earth without walking around the sun.

1) Which object seems to have the most simple orbit motion?

2) Which object seems to have the most complex motion?

3) Which object traveled the longest distance?

Part 2: Give students one **long rubber band** and one **small rubber band**. Have them choose three small objects: one to represent the sun, one to represent the Earth, and one to represent the moon. Have them build a model representing the Sun, Earth, moon, and their orbits.

1) What force is causing the Earth to orbit the sun?

2) Is it the same type of force that causes the moon to orbit the Earth? Explain why.

3) Is the moon orbiting the sun? Explain why.

Weather Around the World

Directions:

Have students use the **internet** to take the temperature, wind direction, and precipitation in different cities worldwide. Each student will take data on one city. Compare the data of cities near each other and cities in different regions of the world in the same time frames. Fill in the Data Table 1 below for your city, then fill in Graph 1 for all the cities each student took data on. Use a different color for each city.

Data Table 1

Date	High Temp	Low Temp	Wind Direction	Precipitation

Graph 1

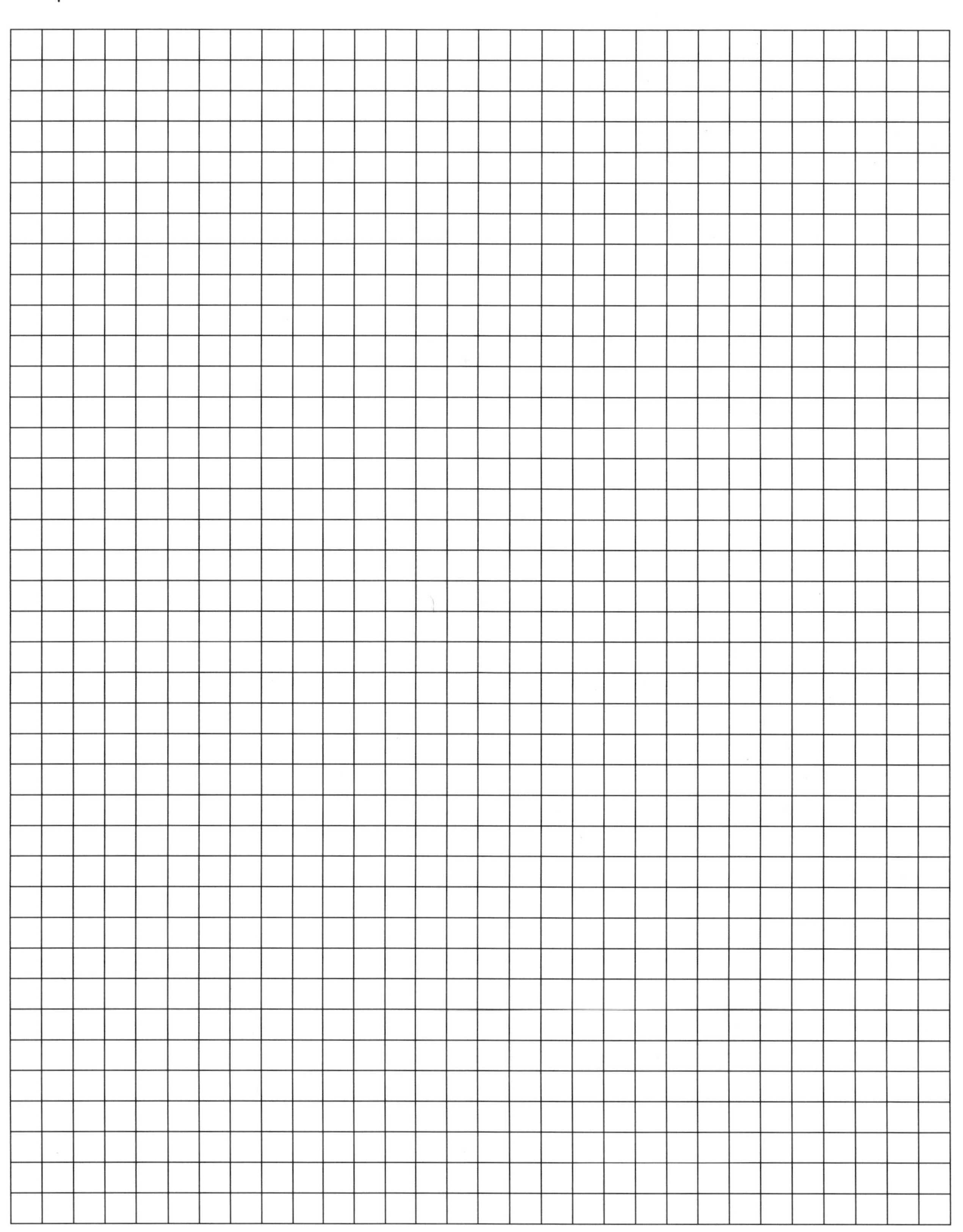

1) Which city did you take data on?

2) What trends do you see in cities in the same hemisphere?

3) What trends do you see in cities in a different hemisphere?

4) What trends did you notice about cities near the equator?

5) What trends did you notice about cities closer to the poles?

6) What trends did you notice about coastal cities?

7) What trends did you notice about cities far from coastlines?

Types of Soil

Directions:

Use the **internet** to research the seven different soil types and describe what they consist of and how they were formed.

1) Type: Silt
    a. It consists of:

    b. It formed from:

2) Type: Clay
    a. It consists of:

    b. It formed from:

3) Type: Sand
    a. It consists of:

    b. It formed from:

4) Type: Loam
    a. It consists of:

    b. It formed from:

5) Type: Peat
    a. It consists of:

    b. It formed from:

6) Type: Chalk
    a. It consists of:

    b. It formed from:

7) Type: Saline
    a. It consists of:

    b. It formed from:

Earthquake Model

Directions:

Construction of the Model: Take a **1x6 piece of wood** at least 3 feet long and **glue 120 grid sandpaper** on its surface. Take a 6-inch piece of **wooden 2x4**, glue 120 grit sandpaper to the bottom, and screw one **screw eyelet** into both ends of the wooden 2x4. Attach a **long thin spring** to the front eyelet on the small wooden 2x4. Build another **wooden 2x4** like the first but with only one **screw eyelet**. Connect the other **spring** to this wooden 2x4. It is best to have enough materials for each group of students as the teacher talks the students through modeling the earthquake.

Earthquake Modeling: Sections of the Earth build up lots of tension over years before they suddenly move; this movement of one surface over another is what causes the earthquake. Place the small 2x4 on the long 1x6 with the sandpapers facing each other. Pull on the spring and watch it stretch (in real life, this takes many years), building up tension; then, the 2x4 will suddenly move; this is the earthquake.

Aftershocks: You can attach the second block to the first, hooking the second spring between the two 2x4s, and see how aftershocks happen. Pull on the first spring; when the first block moves, this puts tension on the second spring priming the second block to later move, causing an aftershock.

Effects on Surface Objects: Add a **small tower**, a **little army man**, or some other figure on top of a 2x4 to show how earthquakes can knock things over, potentially causing lots of damage.

1) Have the students describe the changes in the Earth's surface from an Earthquake.

Landslide Model

Directions:

Place a few inches of dry **dirt** or **sand** inside a **large tub**. Place the tub on a table and lift up on one side of the tub until the dirt/sand starts to move. Then tap the bottom of the tub to cause a section of dirt/sand to move. Landslides are caused by a disturbance in the stability of a slope. Causes could be heavy rain, drought, earthquake, volcanic eruption, or human activities.

1) Have the students describe the changes in the Earth's surface from a landslide.

Volcanic Eruptions

Directions:

Effusive Eruption: Lava flowing out of a volcano can be easily modeled with a **Play-Doh** mound creating a hole by pushing a finger through the thickest surface. Place some **baking soda** with **red food coloring** at the bottom of the hole. Add some **vinegar** to the baking soda to make the lava flow out of the hole.

1) How does the effusive volcanic eruption change the Earth's surface when the lava cools into rock?

2) What does this produce?

Explosive Eruption: Gas pressure that shatters rock and thick magma get shot up into the atmosphere; it can cause massive damage quickly. This model is more dangerous to show. Everyone needs to wear **safety goggles** and **lab aprons**.

1) Take **Play-Doh** out of its jar that is a different color (so it is easy to see later) from your table/desk and floor and make it into a thin sheet.
2) Attach a **small balloon** to an **air pump** (you may need a **rubber band** tightly wound around the nozzle to seal the balloon).
3) Place the balloon under the thin sheet of Play-Doh. Pump up the balloon until it explodes.
4) Look to see how far away pieces of the Play-Doh and balloon flew. The balloon would be falling magma, and the Play-Doh would be the Earth's crust.
5) Describe how the Earth's surface changes when an explosive volcanic eruption occurs.

6) Which eruption seems to be the most dangerous? Explain why.

Reducing Weather-Related Hazards

Directions:

Have students look on the **internet** to investigate claims that builders can build homes with weather-resistant features. Have students explain how the technology of their products allows companies to make these claims.

1) Hurricane-proof houses:

2) Wind-resistant roofs:

3) Hail-resistant roofs:

Different government agencies have spent millions of dollars on flood prevention. Investigate how governments have built structures to reduce the impact of

4) Coastal storm surge:

5) Watershed drainage:

Notes:

Natural and Manmade Resources

Directions:

Natural resources come directly from nature to support life and meet peoples' needs. **Manmade resources** are when humans take those natural resources to produce something new that adds value to our lives. Use these definitions to cut out and divide the resources listed below into two piles **1) Natural Resources** and **2) Manmade Resources**.

Resources List

- Agriculture
- Air
- Air conditioners
- Airplanes
- Animals
- Birds
- Bridges
- Cement
- Cities
- Coal
- Electricity
- Factories
- Fertile land
- Fish
- Forests
- Harbors
- Human population
- Machinery
- Metals
- Minerals
- Money
- Natural gas
- Oil/petroleum
- Paper
- Phones
- Plants
- Plastic
- Roads
- Rubber
- Sand
- Soil
- Stone
- Sunlight
- Tools
- Vehicles
- Watches
- Water

This page will be cut from the previous page.

1) Take the pile of natural resources, determine which resources are used in construction, list them here, and explain how each is used.

2) Take the pile of natural resources, determine which resources are used in agriculture, list them here, and explain how each is used.

3) Take the pile of natural resources, determine which resources are used in transportation, list them here, and explain how each is used.

4) Take the pile of natural resources, determine which resources are used in making products, list them here, and explain how each is used.

5) Why is it important to conserve natural resources?

6) What are ways we can reduce the use of natural resources? Give some examples.

7) How would reusing products help conserve natural resources?

8) How does recycling help conserve natural resources?

Water to Ice

Directions:

Take a **plastic bottle of water** half full and place a mark on the bottle showing the height of the water in the bottle. Place it in the **freezer** to determine what happens to the water when it freezes.

1) What happened to the volume of water when it froze?

2) What effect could this have on anything it was contained in?

3) Why do we need to insulate our pipes in the winter?

4) Place ice in liquid water; what does it do? Explain why.

Changes in Food Chains

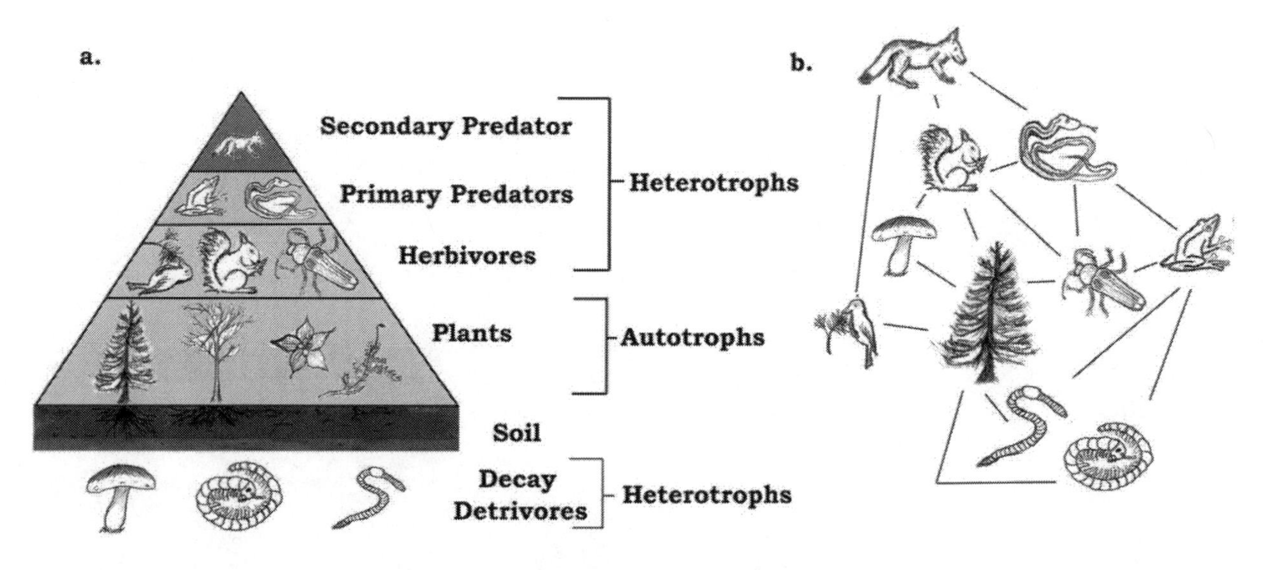

Diagram from Thompsma

Directions:

Use the diagram above to help answer the questions below.

1) Place arrows on the lines showing who eats who in diagram b.
2) If the frogs were to be removed from here, how would this affect the organisms it eats?

    a. How would it affect the organisms that eat the frogs?

3) If the wolf was removed, how would that affect the squirrel and bird populations?

    a. How could that affect the trees in the forest?

4) Which level takes energy from the sun to produce food?
5) Which levels eat food?

6) Before this ecosystem goes into winter, what will happen to the leaves on the tree? Discuss with your teacher why this needs to happen.

7) How would winter affect the amount of food available to the ecosystem?

8) Why would frogs and snakes burrow underground before winter sets in?

9) What do many birds do at this time? Explain why.

10) What could happen to this ecosystem if there was a severe drought in this ecosystem?

    a. What would this do to the animals that live there?

11) What would happen if this area permanently flooded?

Life's Changing Past 1

Directions:

Use the chart of the fossil record below to answer the questions that follow.

mya = Million years ago

Fossil Record Chart

	Period Name	Age of Fossils	Life Found in the Fossils
Cenozoic Era Mammals & Birds	Quaternary	Now-2.4 mya	Cycles of glacial growth and retreat, extinction of many large mammals and birds, Hominids spread
	Tertiary	2.4-65.5 mya	Mammals diversify rapidly
Mesozoic Era Reptiles dominate	Cretaceous	65.5-145 mya	Dinosaurs dominated, then died 65.5 mya
	Jurassic	145-200 mya	Many new dinosaurs emerged
	Triassic	200-251 mya	Dry climate, Pangaea is mostly desert, the evolution of dinosaurs, appearance of first mammals, gymnosperms survived and mosses and ferns on coastlines
Paleozoic Era Fish dominate	Permian	251-299 mya	Reptiles relacing amphibians on land, gymnosperms, mosses, and ferns are popular plants
	Pennsylvanian	299-318 mya	Widespread swamps, the first true forests appeared (gymnosperms), land animals were still amphibious, and reptiles appeared
	Mississippian	318-359 mya	Plants diversifying, giant horsetails and tree ferns on land, amphibious animals mostly on land
	Devonian	359-397 mya	amphibians appeared on land
	Silurian	397-423 mya	Invertebrates dominant animals; first jawed fish appear, first boney fish appear, and insects appear on land by flying from water
	Ordovician	423-488 mya	Lots of marine invertebrates, red and green algae, and primitive plants appear on land
	Cambrian	488-542 mya	An explosion of multicellular life due to lots of oxygen in the atmosphere, first plants, invertebrates, and vertebrates (jawless fish)
Precabrian Era Microscopic life	Proterozoic Eon	542-2500 mya	Eukaryotes appeared/ first protists, lots of bacteria and algae, oxygen levels rose a little, first multicellular plants appeared in the water
	Archean Eon	2500-4000 mya	First life appears, only anaerobic cyanobacteria, photosynthesis evolved in water
	Haden Eon	4000-5000 mya	No life; Earth covered in water after it cooled

1) What was the first life to appear on Earth?

2) When did the first fish appear?

3) What type of life first appeared on land?

4) When did the first amphibians appear?

5) When did reptiles appear?

6) When did dinosaurs die off?

7) When did the first real forests appear?

8) What happened to life on Earth as time went on?

9) What happened to the environments on Earth as time went on?

10) When did humans appear as the last hominids?

    a. Were there any dinosaurs then?

11) How did the abundance of oxygen seem to affect the development of life?

Modeling Earth Layers

Directions:

Find various **colors of granulated material** (like colored sand). If you want your model to be eatable, you can use baking ingredients, **ground-up crackers**, and **cookies** of different colors. Take a **beaker** of one of your substances and pour it in until it is about ½ inch thick. Take your next material and add a layer about ½ inch thick. Keep adding different colored layers until your beaker is full. Use the model you create to answer the following questions.

1) Where in the beaker is the layer you put in first?

2) Where in the beaker is the layer you put in last?

3) Which layer in your beaker represents the oldest rock layer?

    a. Where would we find the oldest fossils?

4) Where is the youngest rock?

    a. Where would you find the youngest fossils?

5) If each layer contains only certain fossils found in that layer, what can we say about the organisms that made those fossils?

6) Human fossils have never been found with dinosaur fossils; what do you think this tells us?

Animal Adaptations

Directions:

Look at the pictures of the animals below and explain how their external structures function to help them survive in their environment.

Comparing Life Cycles

Directions:

Use the diagrams of these life cycles to help explain how they are alike and different.

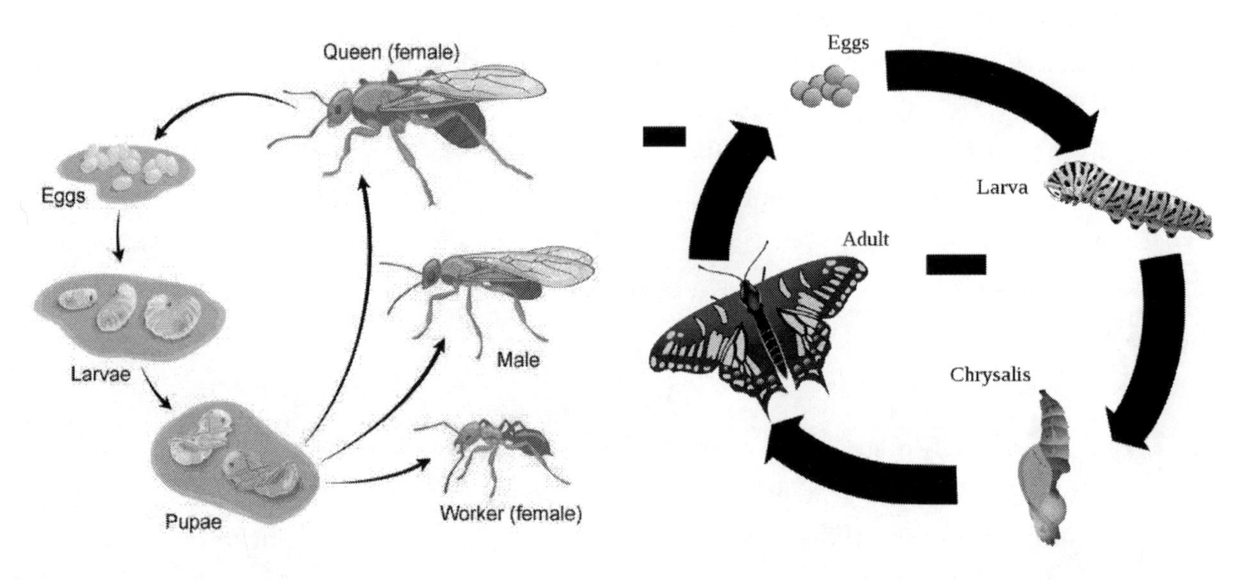

The diagram above is from Tate Holbrook

The diagram above is from Bugboy52.40

1) How are these life cycles alike?

2) How are these life cycles different?

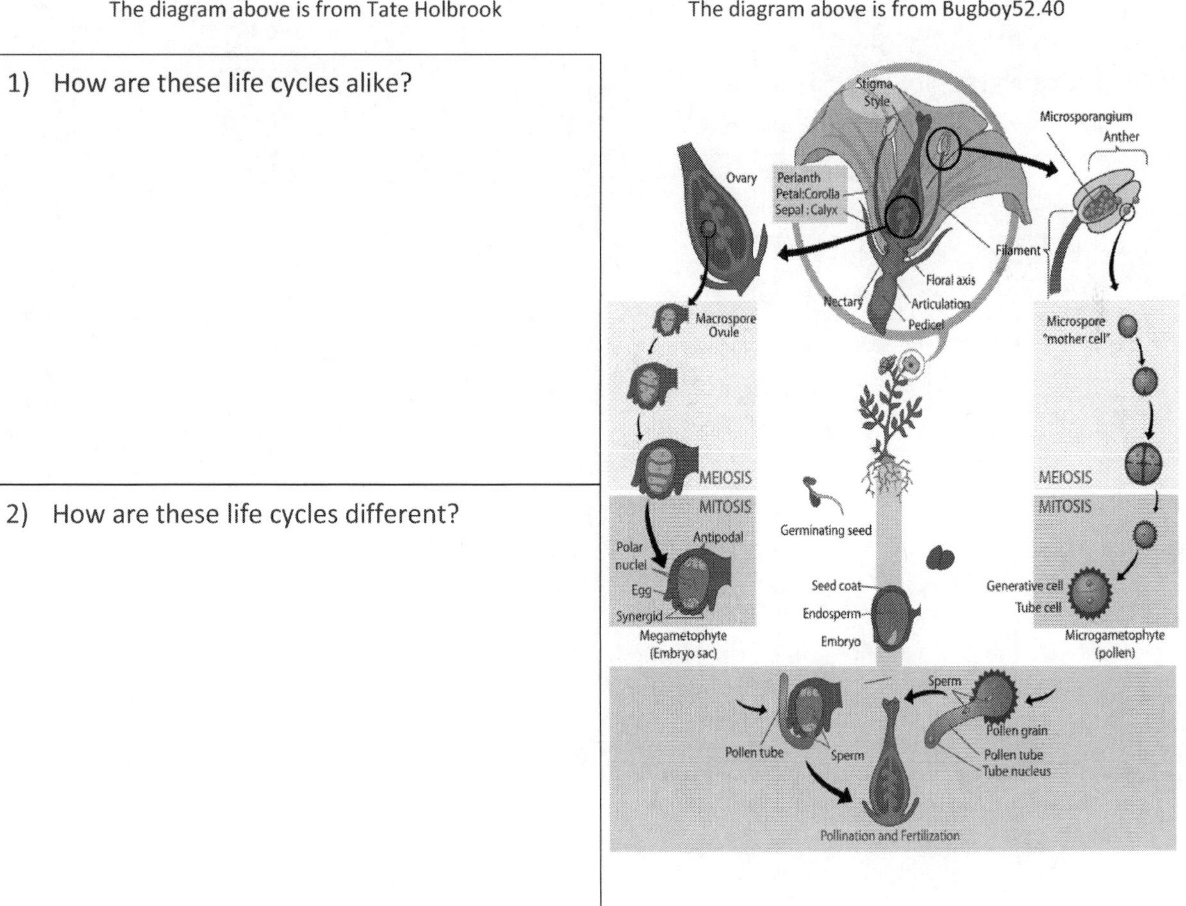

Variation in a Population

Directions:

Each student will need ten **lima beans** and a **metric ruler**. The more students you have, the bigger your data set and the better results you will get.

1) Randomly take ten lima beans out of a bag. Do not use the ones that are broken. Measure the length in millimeters and write this data in Data Table 1.
2) Find the shortest lima bean and the longest lima bean in the class. Then fill in the equal increments between those measurements to make 14 different groups. Then take a class count of how many lima beans fit in each category. Write this data in Data Table 2.
3) Measure the length of your shoe in centimeters. You might want to total your data with the other 3rd Grade classes. What is the length of your shoe?
4) Find the measurement of the shortest shoe of all the classes and the longest shoe of all the classes. Then fill in the equal increments between those measurements to make up 14 different groups. Then take a class count of how many shoes fit in each category. Write this data in Data Table 3. Then fill in the shoe data and let the students copy the count for all classes.
5) Make a graph of the class counts of lima beans data on Graph 1.
6) Make a graph of shoe length data on Graph 2.

Data Table 1

	1	2	3	4	5	6	7	8	9	10
Lima Bean Length (mm)										

Data Table 2

Lima Bean Length (mm)														
Class Count														

Data Table 3

Shoe Length (cm)														
Class Count														
All-Classes Count														

Graph 1

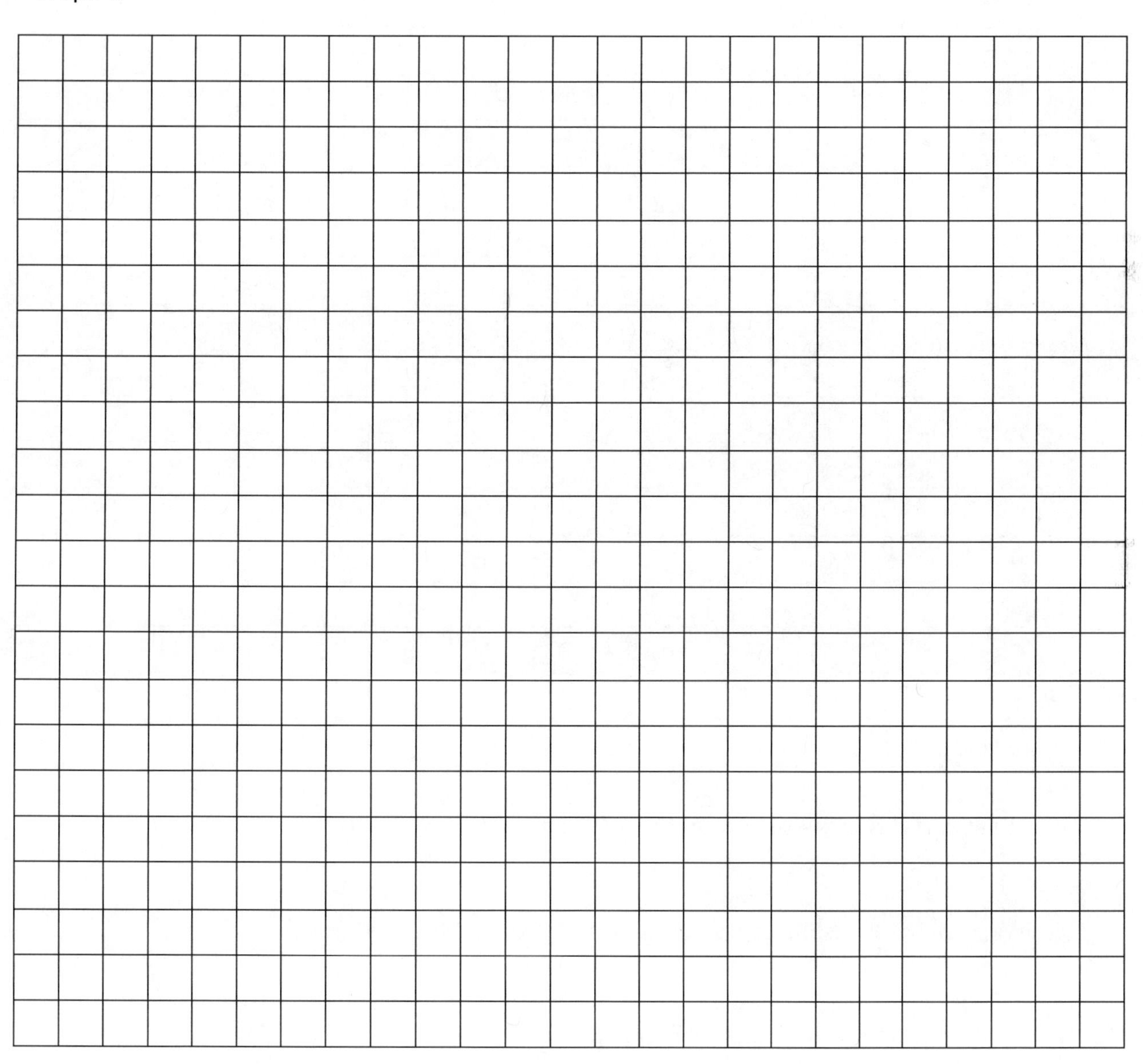

Graph 2

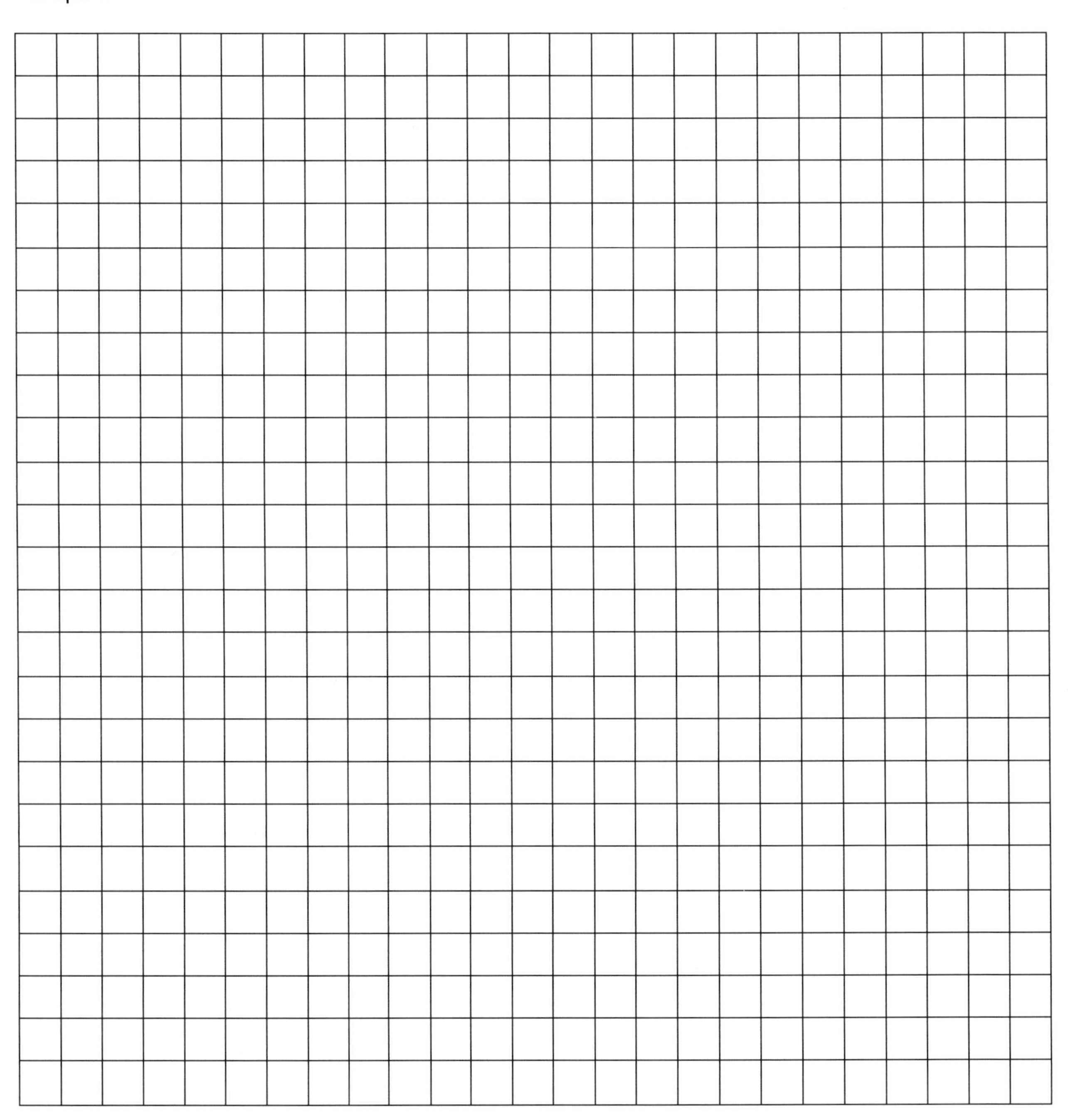

1) What was the shape of each of those graphs?

2) Were all the measurements the same for all the beans and shoes?

3) Why do you think populations have variation?

4) How can this be an advantage?

5) In natural selection, a certain trait is selected by nature to fit best in the environment. How can having a variety help a population survive when there is a sudden environmental change?

6) Will the population look the same 1000 years after the environment's change selects new successful individuals? (Explain)

7) How could long toes become an advantage?

8) How could long toes be a disadvantage?

Changing Environments for Beads

Directions:

You will need **red**, **white**, and **blue beads** in a **bowl**, **red**, **white**, and **blue construction paper**, and **colored pencils.**

1) Have each group randomly get 10 beads out of the bowl and place them on white paper. The beads will represent a population with three variants, and the paper will represent the environment they are in. Count the number of beads of each color and write this in Data Table 1 for the first generation.
2) The students will represent a predator of the beads. Have the students in each group take out three beads that do not match the environment's background, place them back into the bowl, and randomly pick three more beads out of the bowl (if you do not have any that don't match the background take ones that do match until you have three).
3) Add the three new beads to the paper, count how many beads there are for each color in the population, and write this down in Data Table 1 for Generation 2.
4) Repeat steps 2 and 3 for six more generations.
5) After completing eight generations change the white background to red or blue and predict how your population will change over time.
6) Repeat steps 2 and 3 for eight generations and write this in Data Table 2.
7) Once you have completed both Data Tables, graph your data for Data Table 1 on Graph 1 and Data Table 2 on Graph 2, making a line graph using red (for red beads), black (for white beads), and blue (for blue beads) colored pencils.
8) Once the graphs are completed, answer the questions that follow.

Data Table 1

Bead Color	Gen 1	Gen 2	Gens 3	Gen 4	Gen 5	Gen 6	Gen 7	Gen 8
Red								
White								
Blue								

Data Table 2

Bead Color	Gen 1	Gen 2	Gen 3	Gen 4	Gen 5	Gen 6	Gen 7	Gen 8
Red								
White								
Blue								

Graph 1 (Line Graph) Color of Background: White

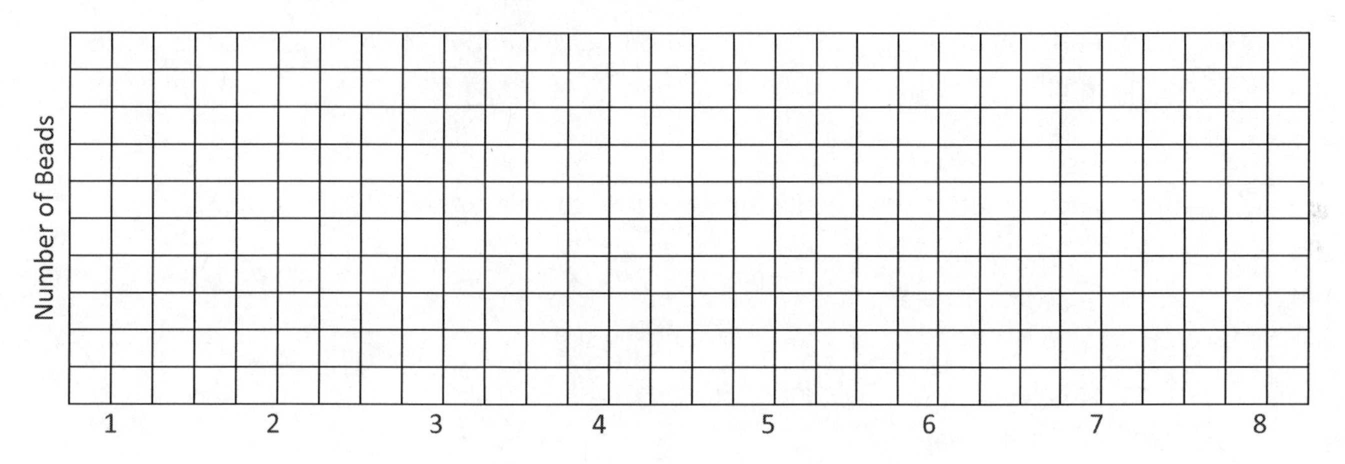

Graph 2 (Line Graph) Color of Background: ___________________________

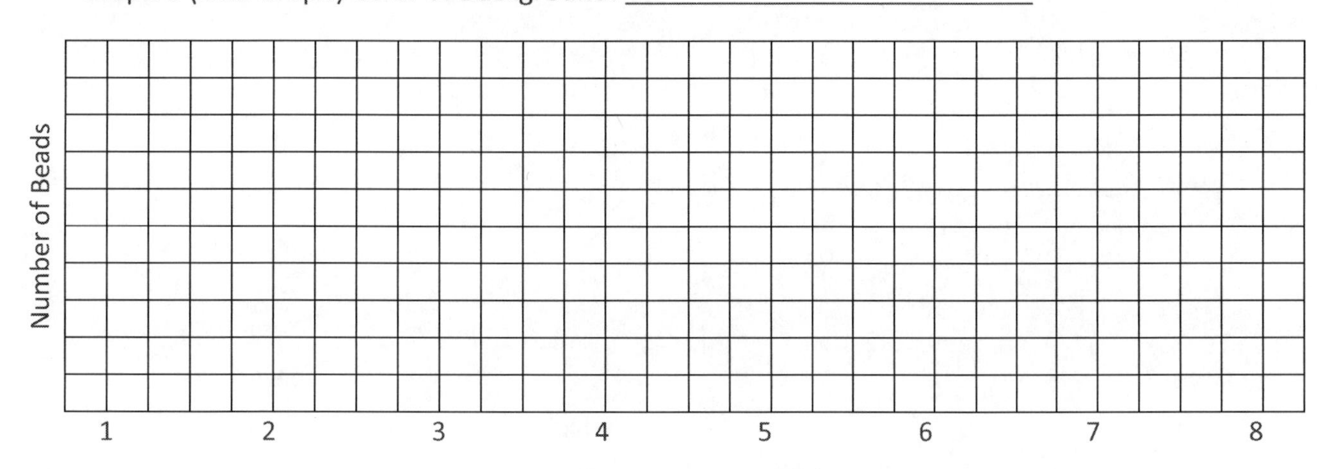

Questions:

1) How did the population of beads change when there was a white background?

2) Why would this be useful in nature?

3) How did the population of beads change when there was a red or blue background?

4) Why would variations in a population benefit a population when an environment changes?

5) How could having variations in a population be hazardous to some individuals in the population?

6) What could happen to this population if the background turned green?

7) Which population would be more fit, one with little variation or one with lots of variation? Explain why.

Goldfish Evolution

Directions:

You will need **food serving gloves** for the teacher, a **large mixing bowl**, **paper plates**, **cheese-flavored Goldfish Crackers**, and **pretzel-flavored Goldfish Crackers**. In this activity, students will represent predators, a goldfish-eating shark, which selectively preys upon goldfish in small populations. This shark likes to eat two kinds of fish: yellow fish (cheese-flavored) and brown fish (pretzel flavored). The yellow fish are easy for you to see, so they are easy to catch and eat. Brown fish travel more quickly and can evade capture more easily. Because of this, you eat only yellow fish, unless there are no yellow fish around, in which case you eat the brown fish. Fish are replaced with individuals randomly selected from an ocean (mixing bowl full of Goldfish crackers). Brown fish is determined by the presence of a dominant allele (B), and yellow fish by a recessive allele (b).

1) Send one student from your group with a paper plate to collect a random population of 10 fish (crackers) from the mixing bowl (ocean). Your teacher will place them on your plate for you.
2) In data table 1, for generation 1, record the number of yellow and brown fish in the population.
3) Choose three yellow fish from the population and eat them. If you do not have any yellow fish, fill in the missing number by eating the brown fish for a total of 3 fish eaten.
4) Send one student to the bowl (ocean) to get three more random fish and add them to your population.
5) In data table 1, for generation 2, record the number of yellow and brown fish.
6) Repeat steps 3-5 until you have data for all five generations.

Data Table 1

Generations	#of Gold Fish	# of Brown Fish	% of Gold Fish	% of Brown Fish
1				
2				
3				
4				
5				

7) Using the information from Data Table 1 above, plot your data on Graph 1 to show how your population changed over time. For each generation, plot two separated bars: use one color to represent the percent population of goldfish, and use a different color to plot the percent population of brown fish.

Graph 1

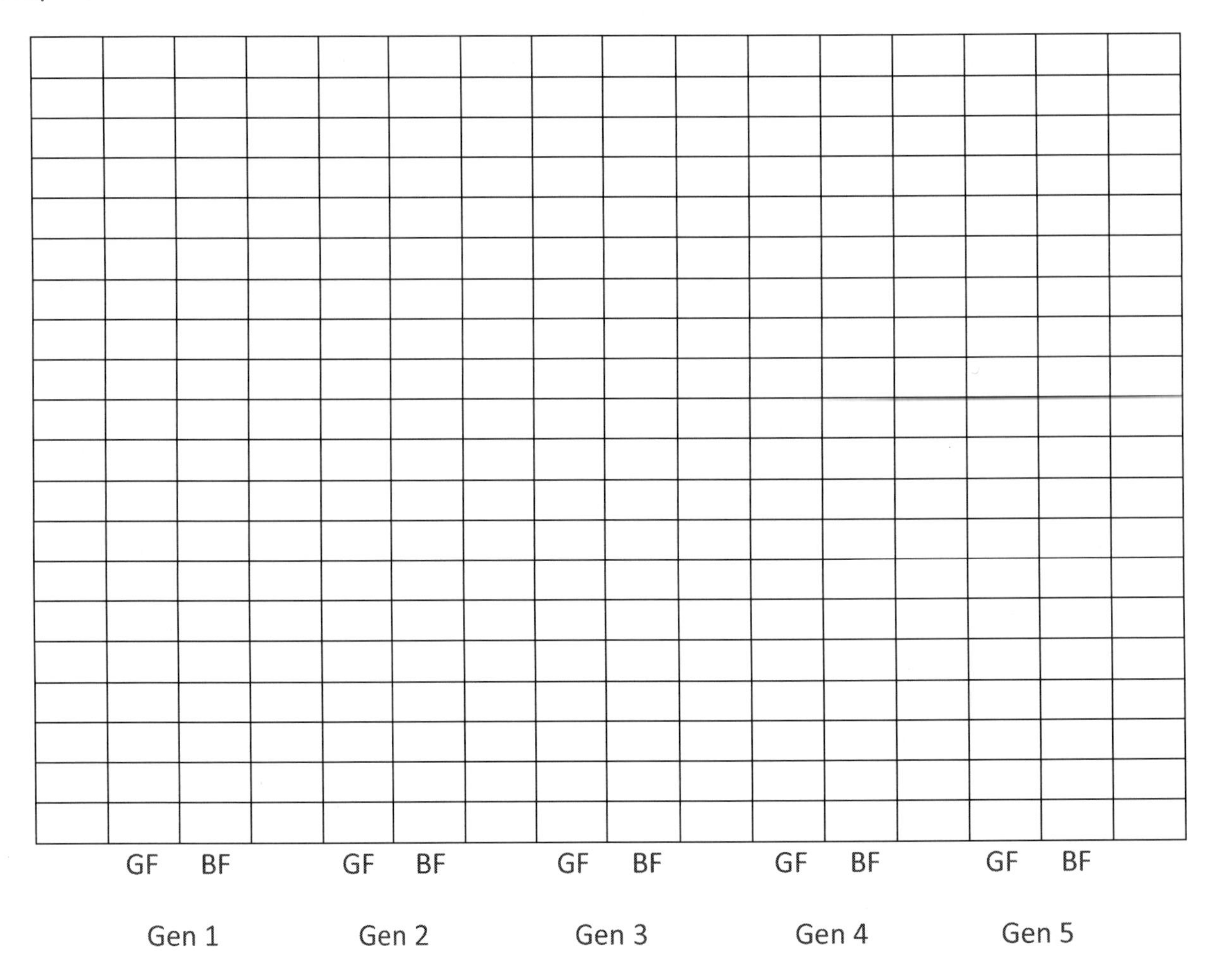

Questions:

1) How did the number of yellow fish change from generation 1 to 5?

2) Which trait was reduced in this population over time? Why?

3) What event occurs if there is a change in a population over time?

4) Explain what would happen over time if the brown fish were easier to catch?

5) What would happen if both fish were equally easy to catch?

How can the Environment Influence these Traits?

1) A person with a sunburn

2) A dog with no tail

3) An overweight dog

4) Dried up plants

5) An adult with no hair on their legs

Social Behavior

Directions:

Use the **internet** to research social behaviors and answer the following questions.

1) Why do birds flock, fish school, and cattle herd? What benefits do these behaviors give them?

    a. How does it help them defend themselves?

2) Why is it beneficial for animals of the same group to cooperate while hunting? Give examples.

3) Why is it beneficial for animals of the same group to cooperate while migrating? Give examples.

4) How does being in a group help animals cope with changes?

Virtual Investigations that go with 3rd Grade Science

ExploreLearning.com

Measuring Volume

Weight and Mass

Density

Mineral Identification

Phases of Water

Energy Conversions

Radiation

Phases of the Moon

Solar System

Prairie Ecosystem

Rabbit Population by Season

Honeybee Hive

Animal Group Behavior

PhET.colorado.edu

Density

Density via Comparison

Determine Density via Water Displacement

States of Matter: Basics

Force and Motion: Basics

Balancing Act

Energy Skate Park: Basics

Energy Skate Park

Energy Forms and Changes

Projectile Motion

Gravity and Orbits

Pendulum Lab

Charges and Fields

Balloons and Static Electricity

Natural Selection

Physics Classroom.com

Force

Rocket Sledder

Skydiving

Kinetic Energy

It's All Uphill

Orbital Motion

Gravitation

3rd Grade TEKS and NGSS Correlations

What Causes Sinking and Floating? Science, 3rd Grade (b) 1ABCDEF 2BCD 3AB 5ABC 6A; 3-PS2-1

Build a Useful Boat Science, 3rd Grade (b) 1ABCDEFG 2ABD 4A 5ABCDFG 6A; 3-PS2-1, 3-5-ETS1-123

Solids, Liquids, and Gases Science, 3rd Grade (b) 1ABCDE 2B 3AB 5AB 6ABC

Magic Rings Science, 3rd Grade (b) 1ABCDE 2B 5ABDE 6A 7A; 3-PS2-34

Condensation of Water Science, 3rd Grade (b) 1ABCDEG 2AB 3ABC 5ABE 6AC 7A; 3-PS2-4

Rolling Brick Tower Science, 3rd Grade (b) 1ABCDEG 2ABCD 3ABC 4A 5ABCDEFG 6AD 7AB; 3-PS2-1, 3-ESS3-1, 3-5-ETS1-123

Let's Swing Science, 3rd Grade (b) 1ABCDE 3ABC 5ABCDE 7AB; 3-PS2-1

Let's Swing Again Science, 3rd Grade (b) 1ABCDE 3ABC 5ABCDE 7AB 8AB; 3-PS2-12

Bouncing Ball Science, 3rd Grade (b) 1ABCDE 3ABC 5ABCDE 7AB 8AB; 3-PS2-12

Types of Energy Science, 3rd Grade (b) 1ABE 3ABC 5AE 8A

Map of our Solar System Science, 3rd Grade (b) 1ABEFG 2AB 4B 5ACD 9AB; 3-PS2-2

How Earth and Moon Orbit the Sun Science, 3rd Grade (b) 1ABCDEF 2AB 3ABC 5ACD 7A 9A; 3-PS2-2

Weather Around the World Science, 3rd Grade (b) 1ABEF 2BC 3AB 4B 5ABEG 10A; 3ESS2-12

Types of Soil Science, 3rd Grade (b) 1ABE 3B 4B 5BEFG 10B

Earthquake Model Science, 3rd Grade (b) 1ABCDEG 2AB 3B 5ABCDEFG 8A 10C

Landslide Model Science, 3rd Grade (b) 1ABCDEG 2AB 3B 5ABCDEFG 8A 10C

Volcanic Eruptions Science, 3rd Grade (b) 1ABCDEG 2AB 3ABC 5ABCDEFG 8A 10C

Reducing Weather Related Hazards Science, 3rd Grade (b) 1AB 3AB 4AB 5F 8A 11AC; 3-ESS3-1

Natural and Manmade Resources Science, 3rd Grade (b) 1ABE 3ABC 4A 5DF 11ABC

Water to Ice Science, 3rd Grade (b) 1ABCDE 2AB 3AB 5ABCG 6C

Changes in Food Chains Science, 3rd Grade (b) 1ABEG 2B 3ABC 5BDEG 12ABC; 3-LS4-34

Life's Changing Past 1 Science, 3rd Grade (b) 1ABEG 2BC 4A 5ABCG 12D; 3-LS4-124

Modeling Earth Layers Science, 3rd Grade (b) 1ABCDEG 2AB 3ABC 5ABCDG 12D; 3-LS4-134

Animal Adaptations Science, 3rd Grade (b) 1ABEG 3ABC 5F 13A; 3-LS4-3

Comparing Life Cycles Science, 3rd Grade (b) 1ABEG 3ABC 5ABDG 13B; 3-LS1-1

Variation in a Population Science, 3rd Grade (b) 1ABCDEF 2BC 3ABC 5ACG 12C 13A; 3-LS3-1, 3-LS4-2

Changing Environments for Beads Science, 3rd Grade (b) 1ABCDEF 2BC 3ABC 5ACG 12C 13A; 3-LS3-1, 3-LS4-234

Goldfish Evolution Science, 3rd Grade (b) 1ABCDEF 2BC 3ABC 5ACG 12C 13A; 3-LS3-1, 3-LS4-234

How can the Environment Influence these Traits? Science, 3rd Grade (b) 1A 3ABC 5BDEG 8A; 3-LS3-2

Social Behavior Science, 3rd Grade (b) 1A 3ABC 4B 5BFG; 3-LS2-1

Equipment List for all 3rd Grade Investigations

If you want to be able to do all the labs in this manual for 3rd Grade Science, here is a list of all the equipment you will need (in order of appearance).

Scales

Rulers

Graduated cylinders

Blocks

Beakers

Water

Oil-based clay

Bucket/tub

Food coloring

Dry ice

Gloves

Erlenmeyer flask

Long golf tee

Glue

Ring magnets

Ice

Balloons

Mega Blocks

Masking tape

Swing set

Rubber balls

Internet

Colored pencils

Variety of rubber bands

1x6 pieces of wood

2x4 pieces of wood

120 grit sandpaper

Screw eyelets

Long thin springs

Large tubs

Dirt

Sand

Play-Doh

Baking soda

Red food coloring

Vinegar

Safety goggles

Air pump

Plastic bottles of water

Freezer

Cookies

Crackers

Lima beans

Red, white, and blue beads

Bowls

Red, white, and blue construction paper

Colored pencils

Food serving gloves

Large mixing bowl

Paper plates

Cheese-flavored Goldfish Crackers

Pretzel-Flavored Goldfish Crackers

4th Grade Science Investigations

Describing Matter in your Classroom

Directions:

Take objects in your room, measure them, and classify them as solid, liquid, or gas. Measure the desk/table with a **meter stick**. Measure the **whiteboard eraser** with a **scale**. Measure the 50 mL of water in a **graduated cylinder** and show the teacher the meniscus is at 50 mL. Use a **thermometer** to measure the temperature of the room. Use a **graduated cylinder** and a **scale** to measure the density of 25 mL of water (make sure to subtract the mass of the graduated cylinder from the water and graduated cylinder). Use a **metric ruler** to measure the volume of a block in cubic centimeters. Fill in Data Table 1 to show your work.

Data Table 1

Object	Type of Measurement	Measurement	Classify
1) Desk/Table	Length	cm	Solid / Liquid / Gas
2) Whiteboard Eraser	Mass	g	Solid / Liquid / Gas
3) Water	Volume	Show teacher 50 mL	Solid / Liquid / Gas
4) Air in Room	Temperature	°C	Solid / Liquid / Gas
5) 25 mL of water	Density	g/mL	Solid / Liquid / Gas
6) Block	Volume	cc	Solid / Liquid / Gas

1) Find the mass of the block in grams to help you find its density. What was the density of the block (Hint: 1cc = 1mL)?

2) Then determine if it would sink or float in water by comparing it to the density of water in Data Table 1. (Hint: 1 mL = 1 cc). Do you think it will sink or float?

3) Then test to see if you were correct. Did the block float or sink?

4) Explain how you determined whether an object was a solid, liquid, or gas.

Mixtures

Directions:

Show examples of these mixtures to the students and have them compare and contrast them.

Salt and water

Brass = copper and zinc

Bronze = copper and tin

Kool-Aid and water

Water and oil

1) How are these mixtures alike?

2) How are these mixtures different?

3) Describe how you could group them.

Conservation of Mixtures

Directions:

Find the mass of three **beakers** of **water**, **salt**, **Kool-Aid**, and **oil** in a **pipette** (subtract the mass of the empty pipette) with a **scale**. Predict the mass of the future mixtures by adding them together. Then mix each together, and find their actual mass. Write all your data in Data Table 1 below.

Data Table 1

Object's Mass	Water and Container Mass	Predicted Mixture's Mass	Actual Mixture's Mass
Salt g +	g =		
Kool-Aid g +	g =		
Oil g +	g =		

1) How did the predicted mass compare to the actual mass?

2) What do you think the word conservation means if the mass of each mixture was conserved?

3) Was mass conserved in these mixtures?

4) What could be a source of error in this activity if your massed did not add up properly?

Making a Compass

Directions:

Take a **bar magnet** and tie it to a **string** in the middle where it will balance. Let it hang; the side that points north is the north side of the magnet. Place an N on the end of the side of the magnet pointing north. Spin the magnet slightly to test if the magnet will always align with the Earth's magnetic field. Write your results in Data Table 1.

Data Table 1

Trial	Direction "N" faced after spin
1	
2	
3	
4	

1) What direction did the magnet face each time?

2) How could you use this knowledge if you were lost?

3) If opposites attract, is the North Pole magnetically north or south?

Comparing Friction Lab

Directions:

You will need an **ice cube**, **rock**, **eraser**, **wooden block**, **aluminum foil**, and a **tray**.

1) Position each object on your tray.
2) Slowly lift one end of the tray and stop when an object slides.
3) Measure the height of the end of the tray you raised.
4) Keep doing this until you have measurements for all of your objects.
5) Put all your data in Data Table 1 below.

Data Table 1

Object	Height at which the object slid (cm)
Ice Cube	
Rock	
Eraser	
Wooden block	
Aluminum foil	

1) Why did the objects slide off at different heights?

2) Which direction is the force of friction relative to movement?

3) Static friction keeps objects from moving; sliding friction is what slows objects as it slides. What type of friction does the object have as it slides down the tray?

4) Friction happens when objects' surfaces slide across each other. Rub your hands together; what do you feel after a while?

    a. What do you hear?

5) What kinds of energy will sliding friction transfer energy to?

Showing Forces

Directions:

Problem: Use a **toy car** or **truck** to investigate how forces cause an object to move in the direction of the force, and the force of friction always acts in the opposite direction.

Hypothesis:

Investigation description:

Results:

Conclusion:

Energy Transformation Balls

Directions:

You will need a pair of **steel energy transformation balls**, an **index card** or piece of **paper**, and **safety goggles.**

Looking at the materials and lab we will be using, what safety precautions should we take to protect ourselves and materials during the investigation?

Put on your safety goggles. Have one person take the steel energy transformation balls and hold one in each hand. Have another person vertically hold out a piece of paper or an index card. The person holding the energy transformation balls should then smash the two balls together on the index card with a very strong force. Make sure not to get anyone's fingers in the way.

1) What do you see on the paper where the balls hit?

2) What do you smell?

3) What do you think happened?

4) How was energy transformed?

Where did the Energy Go?

Directions:

Take a large **kickball** and let it bounce until it stops moving. Answer the questions below to help discover where the energy went. Repeat if necessary.

1) What kind of energy did the ball have when held before it was dropped?
2) What energy did the ball gain as it dropped through the air?
    a. What force caused this to happen?
3) What force was moving against the ball as it dropped?
    a. What did that force cause energy to transfer to?
4) When the ball hit the ground, how did the energy change as the ball changed shape?
5) What did you hear as the ball collided with the ground?
    a. How was energy lost in the collision?
6) As the ball goes up, what forces are acting on the ball?
7) Why does the ball not bounce as high as it was before on each bounce?
8) What evidence showed you all the energy eventually left the ball?
9) Where did the energy go?

Water Waves of Energy

Directions:

Pour water into a **small swimming pool** until it is about half full. Place a **rubber ducky** at the edge of the pool. Drop **balls** of different masses into the pool. Observe what happens and answer the questions below.

Data Table 1

Objects	Mass	Rank the duck movement
Ball 1	g	Most / Middle / Least
Ball 2	g	Most / Middle / Least
Ball 3	g	Most / Middle / Least

1) What happened to the rubber ducky when the balls were dropped into the water?

2) How did the energy transfer from the ball to the rubber ducky?

3) Where did the ball get its energy?

4) Which ball had the most energy when it hit the water?

    a. How could you tell?

5) When left alone, the water in the pool stops moving. Where do you think the energy goes?

Observing Waves in a Slinky

Directions:

You will need a standard to long **slinky** to make both compression waves and transverse waves. You will do this with one person holding one end of the slinky and another person holding the other. One person will move one end; the other will hold still. Each person needs to take a turn moving the slinky to make the waves described below. You will show this to your teacher.

1) Make a compression wave with high frequency/short-wavelength by quickly moving the end of your slinky forward and backward in the same plane as the slinky.
2) Make a compression wave with low frequency/long-wavelength by moving the end of the slinky forward and backward in the same plane as the slinky but slowly.
3) Make a compression wave with high amplitude by repeating the procedure in #2 but making bigger, more violent pushes down the slinky.
4) Make a compression wave with low amplitude, the same as the procedure in #2 but making smaller, less violent pushes down the slinky.
5) Make a transverse wave with high frequency/short-wavelength by quickly moving the slinky's end perpendicular to the slinky.
6) Make a low-frequency transverse wave/long-wavelength by repeating the procedure in #5 but move the slinky more slowly.
7) Make a transverse wave with a high amplitude by repeating the procedure in #5 but moving the end of the slinky to a bigger distance perpendicular to the slinky.
8) Make a transverse wave with low amplitude by repeating the procedure in #5 but not as big a distance as #7.
9) Now make a compression wave with high frequency/short-wavelength and low amplitude.
10) Now make a compression wave with low frequency/long-wavelength and high amplitude.
11) Now make a transverse wave with high frequency/short-wavelength and high amplitude.
12) Now make a transverse wave with low frequency/long-wavelength and low amplitude.

Identifying Conductors and Insulators

Directions:

Set up: You will need a **battery(s)**, a **battery pack** with wires exposed at the end, a **Christmas light** cut, and insulation stripped off the wires' ends. Put the battery(s) into your battery pack. Attach one end of the exposed wire of the battery pack to one exposed end of the Christmas light by twisting them together. This light will be used to see if the materials are conductors or insulators. Make sure it works by taking the free, exposed ends of the battery pack wire and light bulb wire and touching them together. If the bulb lights, it works.

Materials: You will also need a variety of materials like a **penny**, a **wooden spoon**, a **metal spoon** or **fork**, a **paper clip**, **paper**, a **comb**, **aluminum foil**, an **aluminum can** (check both the top of the can and the painted label), a **rubber band**, and a **pencil**.

Looking at the materials and lab we will be using, what safety precautions should we take to protect ourselves and materials during the investigation?

1) Test each of the materials you gathered by taking the exposed free ends of the wires from the battery pack and the Christmas light and touching them both on the material you are testing at the same time on different ends of the material. If the light bulb lights, it is a conductor because electrons can pass through the material. It is an insulator if it does not light because it does not let electrons pass through the material. Fill in Data Table 1 below, listing conductors and insulators.

Data Table 1

The Light bulb lights: **Conductors**	The light bulb does not light: **Insulators**

1) What pattern do you see in the materials that are conductors?

2) What pattern do you see in the materials that are insulators?

3) What other materials could allow the light bulb to light?

4) What other materials might cause the light bulb not to light?

Building a Communication Device

Directions:

Students must design and build a nonelectronic device to communicate sound over a distance. Let students use materials from home or the teacher has available in the classroom. They can develop signals like Morris code or speak. Have the students then share how their devices communicate a message over a long distance to the class. Test them by seeing who can successfully send the most information over the longest distance. Information will be defined as the most letters in the alphabet used in the message.

The Glowing Pickle

Directions:

You need an **extension cord** that has been stripped with **alligator clips** attached to the female end of an extension cord (opposite to the side that has the plug you plug into the wall). Insert two **forks** to opposite ends of a **pickle**. Attach alligator clips onto each fork from the extension cord. Do not plug in the cord until you have inserted the forks into each end of the pickle and the alligator clips are on the forks.

When plugged in: Touching any part of the apparatus (including the pickle) could cause a damaging shock. Please do not touch it until you unplug the set-up. Keep the students away from the circuit, have them only watch.

Turn off the lights and then plug in the cord.

1) How can you tell the circuit is closed?

2) What is happening to the pickle?

3) How can you tell thermal energy is given off?

Conductors in Cooking

1) Conductors transfer heat quickly. Which objects/materials are used to transfer heat to food quickly when we cook?

2) Insulators stop or slow thermal energy moving through them. Which objects/materials keep us from being burned when we cook?

The Light in our Eyes

Directions:

Light is made up of three primary colors: Red, blue, and green. All colors of light can be made from the combinations of these colors. We see colors that are not absorbed by a surface but are reflected to our eyes. The color black absorbs all light and reflects nothing, while white absorbs no colors and reflects all colors of light. When you look at these objects, what combinations of light are being absorbed and which are being reflected by their surfaces.

Colored Object	Colors Absorbed	Colors Reflected
Green grass		
White shirt		
Black pants		
Blue eyes		
Black hair		
Red petals of a rose		
White golf ball		
Purple shirt		
Pink carnation		
Blue jeans		
Mirror		
Sun		
Black hole		

Seasonal Trends

Directions:

Find the data for a year that shows when the sun rises and sets. The longest day of the year is the first day of summer. The shortest day of the year is the start of winter. The start of spring and fall have equal times of day and night. Use this information to predict seasonal trends and answer the questions below.

1) What should happen to the temperature when days are dramatically longer than nights?

2) What should happen to the temperature when the days are dramatically shorter than nights?

3) Which season is the coldest? Explain why.

4) Which season is the warmest? Explain why.

5) Which season consistently gets colder? Explain why.

6) Which season consistently gets warmer? Explain why.

Moon Phases

Directions:

You will need a **ping pong ball** with one half of it **painted black** for each student or group and a **large ball** sitting in the center of the room to represent the sun. The person will be the Earth, and the ping pong ball will be the moon. Have the white part of the moon face the sun (the large ball in the middle of the room) and the dark side of the moon face away from the sun.

1) The diagram below shows how the moon appears with the Earth at each angle at the eight different points shown below.

2) Arrange your ping pong ball (moon) and yourself (Earth) with the sun (large ball in the center of the room) in each of these positions above and shade how each phase of the moon would appear to you from Earth.

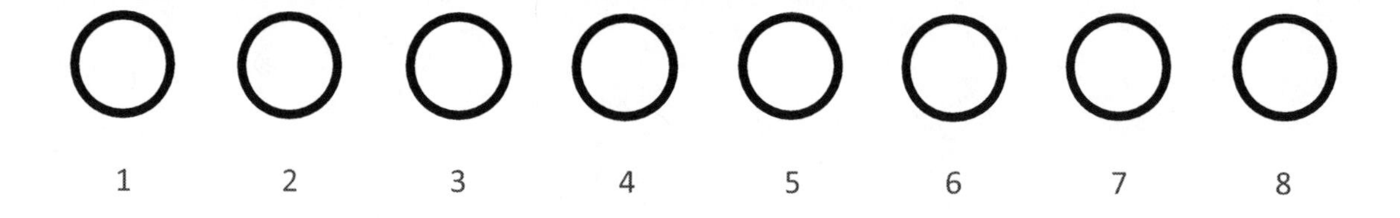

1 2 3 4 5 6 7 8

3) Which phase # is a new moon?

4) Which phase # is a full moon?

5) Which phase # is the first quarter?

6) Which phase # is the third quarter?

7) Explain why we see different phases of the moon.

8) How is this model accurate in showing us the moon phases?

9) How is the model not accurate in showing us the moon phases?

The Water Cycle

Directions:

Use the space below to show how water continually moves over the surface of the Earth. Use pictures, arrows, and the words: evaporation, precipitation, runoff, groundwater, surface water, and transpiration. What role does the sun play in the water cycle?

Weathering, Erosion, and Deposition

Directions:

Use the **internet** to find and show illustrations of how water, wind, and ice slowly change the surface of the Earth. Then describe how each does this.

1) Water:

    a. Describe how water slowly changes the surface of the Earth.

2) Wind:

    a. Describe how wind slowly changes the surface of the Earth.

3) Ice:

    a. Describe how ice slowly changes the surface of the Earth.

4) What careers are going to have an interest in knowing this information?

Weather and Climate

Directions:

Use your observations of the world where you live to help you answer the questions below.

1) Describe the weather for today.

2) How can weather change from day to day?

3) Describe the climate for your area.

4) How does climate change over time for your area?

5) Does the weatherman forecast the climate or the weather? Explain why.

6) Which one is concerned with the state of the atmosphere at a specific place and time?
    a. Weather
    b. Climate
7) Which one is concerned with the long-term characteristics and patterns of weather for the year?
    a. Weather
    b. Climate
8) Which one changes faster?
    a. Weather
    b. Climate
9) Which one changes slowly?
    a. Weather
    b. Climate
10) Describe the difference between weather and climate.

Advantages and Disadvantages

Directions:

Use the **internet** to find the advantages and disadvantages of using these natural resources on the chart.

Resource	Uses	Advantages	Disadvantages
Wind			
Water			
Sunlight			
Plants including leaves, fruit, & wood			
Animals			
Coal			
Oil			
Natural Gas			

The Impact of Energy Resources

Directions:

The different resources in the chart below are all used to make electricity by moving a magnetic field around wires or moving wires through a magnetic field. Use the **internet** or your **textbook** to help explain how each does this and the impact each has on the environment.

Energy Resource	How it makes electricity	Impact on the environment
Coal		
Oil		
Natural Gas		
Nuclear		
Solar		
Wind		
Hydroelectric		
Geothermal		

1) Which resources require conservation to limit the impact on the environment?

2) Which require disposal?

    a. How does the disposal of resources impact the environment?

3) Which resources can be recycled?

    a. How does recycling impact the environment?

Earth Rocks Rock

Directions:

Use the **internet** to help you fill in the chart below for each resource rocks hold and explain their uses and the physical properties that allow them to hold these resources.

Chart 1

Resource in rock	Uses of the resource	Properties that allow rocks to hold this resource
Gypsum, chalk, & slate		
Clay		
Granite & marble		
Salt		
Quartz		
Sulfur		

Resource in rock	Uses of the resource	Properties that allow rocks to hold this resource
Flint		
Garnet		
Talc		
Pumice		
Obsidian		
Copper & zinc		
Iron & aluminum		
Silver & gold		

Resource in rock	Uses of the resource	Properties that allow rocks to hold this resource
Mercury		
Lead		
Limestone, sand, & gravel		
Oil & petroleum		
Coal		
Graphite		

The Gand Canyon

Directions:

Look at the image of the grand canyon. You can see all the layers of rock that have been laid down over billions of years. Sediment forms and stays on the surface as more sediment is laid on top. Over time environments change, and the layers of the rock change colors because of the different chemicals being deposited, slowly locking away the history of the Earth's past until weathering and erosion wear it away like the Colorado River did for the Gand Canyon. Look at the picture of the Gand Canyon below and use that to help you answer the questions that follow.

Picture from Gand Canyon National Park by W. Tyson Joye

1) Where do you see evidence of weathering, erosion, and deposition showing this landscape has changed over time?

2) What do you notice about the layers of rock on different structures?

3) Where are the oldest rocks?

4) Where would you find the oldest fossils?

Modeling Earth Layers

Directions:

Find various **colors of granulated material** (like colored sand). If you want your model to be eatable, you can use baking ingredients, ground-up **crackers**, and **cookies** of different colors. Take a **beaker** of one of your substances and pour it in until it is about ½ inch thick. Take your next material and add a layer about ½ inch thick. Keep adding different colored layers until your beaker is full. Use the model you create to answer the following questions.

1) Where in the beaker is the layer you put in first?

2) Where in the beaker is the layer you put in last?

3) If each layer in your beaker represents a layer of rock, where is the first rock layer that was laid down?

    a. Where would we find the oldest fossils?

4) Where is the youngest rock?

    b. Where would you find the youngest fossils?

5) If each layer contains only certain fossils that are found in that layer, what can we say about the organisms that made those fossils?

6) Human fossils have never been found with dinosaur fossils; what do you think this tells us?

Life's Changing Past 2

Directions:

Use the chart of the fossil record below to answer the questions that follow.

mya = Million years ago

Fossil Record Chart

	Period Name	Age of Fossils	Life Found in the Fossils
Cenozoic Era Mammals & Birds	Quaternary	Now-2.4 mya	Cycles of glacial growth and retreat, extinction of many large mammals and birds, Hominids spread
	Tertiary	2.4-65.5 mya	Mammals diversify rapidly
Mesozoic Era Reptiles dominate	Cretaceous	65.5-145 mya	Dinosaurs dominated, then died 65.5 mya
	Jurassic	145-200 mya	Many new dinosaurs emerged
	Triassic	200-251 mya	Dry climate, Pangaea is mostly desert, the evolution of dinosaurs, appearance of first mammals, gymnosperms survived and mosses and ferns on coastlines
Paleozoic Era Fish dominate	Permian	251-299 mya	Reptiles relacing amphibians on land, gymnosperms, mosses, and ferns are popular plants
	Pennsylvanian	299-318 mya	Widespread swamps, the first true forests appeared (gymnosperms), land animals were still amphibious, and reptiles appeared
	Mississippian	318-359 mya	Plants diversifying, giant horsetails and tree ferns on land, amphibious animals mostly on land
	Devonian	359-397 mya	amphibians appeared on land
	Silurian	397-423 mya	Invertebrates dominant animals; first jawed fish appear, first boney fish appear, and insects appear on land by flying from water
	Ordovician	423-488 mya	Lots of marine invertebrates, red and green algae, and primitive plants appear on land
	Cambrian	488-542 mya	An explosion of multicellular life due to lots of oxygen in the atmosphere, first plants, invertebrates, and vertebrates (jawless fish)
Precabrian Era Microscopic life	Proterozoic Eon	542-2500 mya	Eukaryotes appeared/ first protists, lots of bacteria and algae, oxygen levels rose a little, first multicellular plants appeared in the water
	Archean Eon	2500-4000 mya	First life appears, only anaerobic cyanobacteria, photosynthesis evolved in water
	Haden Eon	4000-5000 mya	No life; Earth covered in water after it cooled

1) What was the first ecosystem?

    a. How do you know?

2) When did life first appear on land?

3) When was the land mostly desert?

4) When was the Earth mostly swampy?

5) List the different types of environments this fossil record shows us.

6) When did the Earth have a series of ice ages?

7) 99.9% of all species that have been on this Earth are now extinct. What happened to life on Earth as time passed?

8) What do you think caused life to change on Earth?

9) Which period did humans appear as the last hominids?

10) What kind of effect are humans having on the Earth's environment?

Topographical Map of the Earth

Directions:

You will need a **topographical map** of Earth to fill out the information below.

1) Where are major mountain ranges located?

    a. Which is the tallest?

2) Approximately how far away is your home from the nearest mountain?

3) Where is the deepest part of the ocean?

4) Approximately how far away is your home from the coast of the closest ocean?

5) What different geographical features can you see on the map of the continent you live on?

    a. What type of biomes do you think are located in these different geographical features? Give evidence of how you know.

Interpreting Topographic & Geological Maps of your State

Directions:

You will need a **topographical map**, a **geological map** of your state, and a **geologic time scale** to fill out the data below.

1) What is your state?

2) Where is the highest elevation in your state?

    a. Describe the surface rocks in that area.

3) Where is the lowest elevation in your state?

    a. Describe the surface rocks in that area.

4) What are your state's different geographical features, and where are they located?

5) What patterns do you see where different types of rock are found?

    a. Why do you think the pattern is there?

Weathering

Directions:

You will need **safety goggles**, **sugar cubes** and **chalk** to represent rocks, **water** in a **beaker**, **vinegar** in another, **pipettes,** and a small **tray/pan**. Weathering is the breakdown of rock by wind, water, and chemical reactions.

Looking at the materials and lab we will be using, what safety precautions should we take to protect ourselves and materials during the investigation?

1) The wind has particles that, when they hit a rock, microscopic pieces break off, making the rock smaller. Rub your hand across a sugar cube while it is over the tray/pan; what does this do to the cube?

2) Take another sugar cube, place it in the tray/pan, and squirt water on it from a pipette. What does this do to the sugar cube?

3) What happens to water when it freezes?

   a. How can this action help break apart rocks?

4) Put on your safety goggles. Take some chalk (representing limestone) and squirt water on it. Now take some vinegar (representing an acid) and squirt that on the chalk. What do you see chemically taking place?

5) How does this lab model weathering?

6) What is not accurate about this model?

Erosion

Directions:

You will need an **apron**, **safety goggles**, **shovel**, **dirt** or **sand** in a **large tub/pan**, a **pitcher** of **water**, a **hairdryer**, a **spray bottle** that sprays **water**, the **internet**, and a **textbook**.

Looking at the materials and lab we will be using, what safety precautions should we take to protect ourselves and materials during the investigation?

1) Go outside, dig up some dirt/sand, and put it into your large tub/pan. Fill it half full. Bring it inside to your lab table.
2) Angle your tub/pan so one end is higher than the other. We are going to simulate **wind** hitting the ground with a hairdryer. Put on your safety goggles. Plug in your hair dryer and keep it away from water. Turn on your hairdryer, have everyone in your group stand behind it so dirt does not blow on them, and bring the hairdryer closer to the dirt in your tub. What do you see happening to the small particles of dirt?

    a. How would plant life rooted in the dirt and covering it affect this action?

    b. You can place your hand in front of the blowing dryer to see. How does putting an object between the wind source and the dirt affect the movement of the dirt?

3) Unplug your hairdryer and set it aside. Take your spray bottle to simulate what a light rain might do to the dirt. Set it to mist and spray the dirt. This model simulates **sheet erosion**. How does the dirt move? Describe the pattern it is making.

4) Now turn on the hairdryer again and aim it at the moist dirt. How does soil with moisture behave differently in the wind than soil without moisture?

5) Adjust your spray bottle to make a stream come out. This model will simulate heavier rain. Aim your bottle at an area of the dirt standing behind the high end of the tray and spray. How is the dirt movement different?

    a. What do you see forming in the dirt?

b. Have multiple spray bottles spray at the same time. This model will simulate **rill erosion**. How is this pattern different than sheet erosion?

c. How do you think plant life rooted in the dirt and covering it affects this action?

6) Now take your pitcher of water to simulate **gully erosion** and pour it from the high end of the tub. How is this dirt movement different from the others?

7) Take your wet dirt/sand back outside and dump it where your teacher tells you to. And dry your tub/pan for the next class.
8) Use the internet and your textbook to fill in the chart below describing the three types of erosion we just modeled. Then research answers to the questions that follow.

Chart 1

Type	Picture	Description
Sheet Erosion		
Rill Erosion		
Gully Erosion		

1) How does the volume of water affect the rate at which erosion happens?

2) How does plant life slow the erosion of soil?

3) Why do most yards have grass on them?

4) What happens when that grass is worn away?

5) Fill in the chart below describing how we slow the erosion rate with each method.

Chart 2

Method	Description
Contour Plowing	
Strip-Cropping	
Terracing	
Crop Rotation	
Windbreaks	

Reducing the Impact of Natural Disasters

Directions:

Divide your class into groups and have each group develop solutions to reduce the impacts of natural disasters on humans. Each group will prepare for a different disaster. The disasters they must prepare for are earthquakes, floods, tsunamis, and volcanic eruptions. Have each group share their solutions with the class when they are finished.

1) What is the disaster you are preparing for?

2) How can we reduce the impact of this natural disaster?

Conservation of Water in Plants

Directions:

Show examples and explain how the plant adaptations below help conserve water.

1) Waxy leaves
2) Closing of the Stoma
3) Deep Roots

Inherited and Acquired Traits

Directions:

- **Inherited traits** appear because of the genetic code in DNA that parents passed down.
- **Acquired traits** appear because of environmental influence.

Use these statements above to help you list traits we have seen in our lives as either inherited or acquired in real life. Have the students look at each other to find traits and then identify those traits as inherited or acquired. Then have students look at plants and other animals and list as many traits as they can see, observe, or remember from the past.

Inherited Traits	Acquired Traits

Photosynthesis and Plant Anatomy

Directions:

The formulas below show plants take in sunlight, water, and carbon dioxide to make food and oxygen. Draw a picture of a plant below and show where the sunlight, water, and carbon dioxide go into the plant, where the food is stored (the parts animals eat), and where the oxygen comes out. Use the **internet** and/or your **textbook** to research how these things enter and leave a plant.

Water + carbon dioxide + sunlight → food + oxygen

$$H_2O + CO_2 + \text{sunlight} \longrightarrow C_6H_{12}O_6 + O_2$$

Plant Structures and Functions 2

Directions:

Find a picture of a plant where you see the roots, stems, leaves, flowers, fruit, and seeds. Print/cut out the picture, **paste** it below, and label its parts. Then next to each labeled plant part, explain how these structures support survival, growth, behavior, and reproduction.

Animal Structures and Functions 2

Directions:

Pick an animal, find a picture, print it, or cut it out and **paste** it on this page. Label the structures of the animal (internal and external) and explain how each is used to support survival, growth, behavior, and reproduction.

The Bodies' Processing and Response

Directions:

Animals receive different types of information through their senses, process the information in their brain, and respond in different ways. Humans are a type of animal that also do these things. Watch people doing each of these actions and tell which senses take in information, where that information is processed, and how the body responds to do that action.

1) Playing soccer

2) Eating food

3) Playing basketball

4) Having a conversation with someone

5) Driving

6) What do your heart and lungs do when you do a lot of exercise?

    a. Why does this happen?

7) When your body gets cold, what does it do?

    a. Why do you think it does this?

8) What does your body do when it gets too hot?

    a. Why do you think this happens?

Food Webs and Pyramids

Directions:

Use the diagram below to help answer the questions.

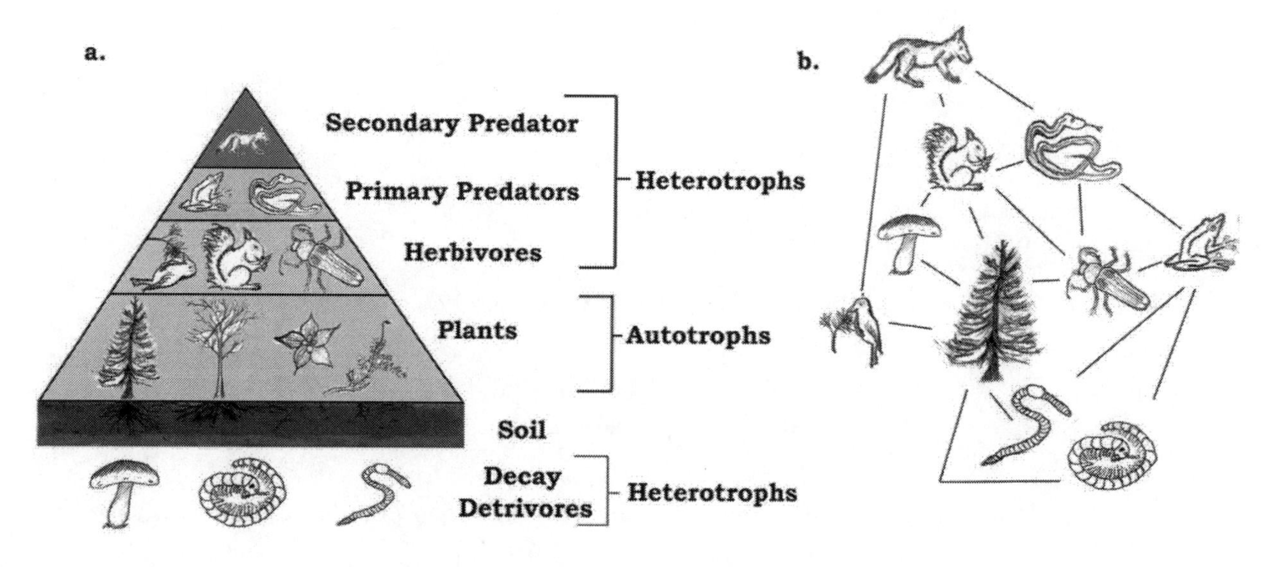

Diagram from Thompsma

1) Put arrowheads on the lines in diagram (b) to show who eats who.
2) What role does the sun play in the community?

    a. What would happen to the community if the sun were to go away?

3) What role do the plants play in this community?

    a. What would happen to this community if the plants were to go away?

4) What role do the Herbivores play in the community?

a. What would happen to this community if the herbivores were to go away?

5) What role do the predators play in the community?

a. What would happen to this community if the predators were to go away?

6) What role do the decomposers (decayers/detrivores) play in the community?

a. What would happen to the community if the decomposers were to go away?

7) How do all of these organisms depend on each other for their survival?

8) What is the ultimate energy source of the material that makes up a carnivore in this or any ecosystem on Earth?

9) What are typical ways humans could disrupt this community?

10) What do you think interdependence means?

Virtual Investigation that go with 4th Grade Science

ExploreLearning.com

Measuring Volume

Weight and Mass

Density

Mineral Identification

Phases of Water

Energy Conversions

Pendulum Clock

Incline Plane Rolling Objects

Radiation

Circuit Builder

Additive Colors

Building Topographical Maps

Weathering

Erosion Rates

Prairie Ecosystem

Comparing Climates (Customary)

Observing Weather (Customary)

Observing Weather (Metric)

Water Cycle

Phases of the Moon

Prairie Ecosystem

Forest Ecosystem

Photosynthesis Lab

Cell Energy Cycle

Inheritance

Germination

PhET.colorado.edu

Density

Concentration

States of Matter: Basics

Greenhouse Effect

Force and Motion: Basics

Balancing Act

Waves Intro

Color Vision

Wave Interference

Pendulum Lab

Energy Skate Park: Basics

Energy Forms and Changes

Gravity and Orbits

Gravity Force Lab: Basics

Circuit Construction Kit: DC-Virtual Lab

Circuit Construction Kit: DC

John Travoltage

Balloons and Static Electricity

Natural Selection

Physicsclassroom.com/Physics-Interactives

Force

Rocket Sledder

Skydiving

Kinetic Energy

It's All Uphill

Orbital Motion

Gravitation

Simple Wave Simulator

DC Circuit Builder

Magnetic Fields

Slinky Lab

RGB Color Addition

4th Grade TEKS and NGSS Correlations

Describing Matter in your Classroom Science, 4th Grade (b) 1ABCDEF 3B 5C 6A

Mixtures Science, 4th Grade (b) 1ABE 3AB 5A 6AB

Conservation of Mixtures Science, 4th Grade (b) 1ABCDEF 2BCD 3ABC 5ABCD 6BC

Making a Compass Science, 4th Grade (b) 1ABCDEFG 2ABD 3ABC 4A 5ADE 6A 7

Comparing Friction Lab Science, 4th Grade (b) 1ABCDEF 2B 3AB 5ABEG 7 8A; 4-PS3-2

Showing Forces Science, 4th Grade (b) 1ABCDE 2ABD 3ABC 5ABDEG 7 8A; 4-PS3-12

Energy Transformation Balls Science, 4th Grade (b) 1ABCDE 2B 3ABC 5ABEG 8A; 4-PS3-123

Where did the Energy Go? Science, 4th Grade (b) 1ABCDE 2B 3ABC 5ABCE 7 8A; 4-PS3-123

Water Waves of Energy Science, 4th Grade (b) 1ABCDEF 2B 3ABC 5ABCDE 7 8A; 4-PS3-12, 4-PS4-1

Observing Waves in a Slinky Science, 4th Grade (b) 1ABCDEG 2B 3B 5ABCDE 8A; 4-PS4-1

Identifying Conductors and Insulators Science, 4th Grade (b) 1ABCDEF 2B 3AB 5ABDE 8BC; 4-PS3-4

Building a Communication Device Science, 4th Grade (b) 1ABCDG 2ABD 3BC 4A 5ABCDE 8A; 4-PS4-3, 3-5-ETS1-123

The Glowing Pickle Science, 4th Grade (b) 1ABCDE 3AB 5ABCDE 8BC; 4-PS3-4

Conductors in Cooking Science, 4th Grade (b) 1AB 3AB 8B

The Light in our Eyes Science, 4th Grade (b) 1ABEF 2B 3AB 5ACE; 4-PS4-2

Seasonal Trends Science, 4th Grade (b) 1ABE 2BC 3ABC 4B 5ABCDEG 9A

Moon Phases Science, 4th Grade (b) 1ABCDEFG 2ABD 3ABC 5ABCDEG 9B

Water Cycle Science, 4th Grade (b) 1ABFG 2AB 3B 4B 5CDEG 10A

Weathering, Erosion, and Deposition Science, 4th Grade (b) 1ABEFG 2B 3ABC 4B 5ABDEG 10B

Weather and Climate Science, 4th Grade (b) 1ABE 3ABC 5ABCDEG 10C

Advantages and Disadvantages Science, 4th Grade (b) 1ABF 3AB 4B 5DEG 11A; 4-ESS3-1

Conservation, Disposal, and Recycling Science, 4th Grade (b) 1ABF 3AB 4B 5DEG 11B; 4-ESS3-1

Earth Rocks Rock Science, 4th Grade (b) 1ABD 3AB 4B 5BD 11C

The Grand Canyon Science, 4th Grade (b) 1ABE 3AB 5ABCDG 11C 12C; 4-ESS1-1, 4-ESS2-1

Modeling the Earth Layers Science, 4th Grade (b) 1ABCDEG 2AB 3ABC 5ABCDG 12C; 4-ESS1-1

Life's Changing Past 2 Science, 4th Grade (b) 1ABG 3ABC 5AC 12C; 4-ESS1-1

Topographical Map of Earth Science, 4th Grade (b) 1ABDG 2AB 3AB 5ACD; 4-ESS2-2

Interpreting Topographical and Geological Maps of your State Science, 4th Grade (b) 1ABDG 2AB 3AB 5ACD; 4-ESS2-2

Weathering Science, 4th Grade (b) 1ABCDEG 2ABD 3ABC 5ABCG 10B; 4-ESS2-1

Erosion Science, 4th Grade (b) 1ABCDEG 2ABD 3ABC 5ABCG 10B 11B; 4-ESS2-1

Reducing the Impact of Natural Disasters Science, 4th Grade (b) 1AB 3ABC 4AB 5ABDE 7 8A; 4-ESS3-2, 3-5-ETS1-1

Conservation of Water in Plants Science, 4th Grade (b) 1ABEFG 3ABC 4B 5ABEFG 13A

Inherited and Acquired Traits Science, 4th Grade (b) 1ABEF 3AB 5BFG 13B

Photosynthesis and Plant Anatomy Science, 4th Grade (b) 1ABEFG 2AB 3ABC 4B 5BCDEF 12A

Plant Structures and Functions 2 Science, 4th Grade (b) 1ABEFG 2AB 3AB 4B 5CDEF 12A; 4-LS1-1

Animal Structures and Functions 2 Science, 4th Grade (b) 1ABEFG 2AB 3AB 4B 5CDEF; 4-LS1-1

Bodies Processing and Response Science, 4th Grade (b) 1ABE 3ABC 5ABDFG; 4-LS1-2

Food Webs and Pyramids Science, 4th Grade (b) 1ABEG 2B 3ABC 5BCDEFG 12B

Equipment List for all 4th Grade Investigations

If you want to be able to do all the labs in this manual for 4th Grade Science, here is a list of all the equipment you will need (in order of appearance).

Meter sticks

Scales

Graduated cylinders

Thermometers

Metric rulers

Salt

Water

Brass

Bronze

Kool-Aid

Oil

Pipettes

Beakers

Bar magnets

String

Ice cubes

Rocks

Erasers

Wooden blocks

Aluminum foil

Trays

Toy cars or trucks

Steel energy transformation balls

Index cards

Paper

Safety goggles

Kickballs

Small swimming pool

Rubber ducky

Selection of balls

Slinkies

Batteries

Battery packs

Christmas lights

Pennies

Wooden spoons

Metal spoons and forks

Paper clips

Combs

Aluminum cans

Rubber bands

Pencils

Extension cord

Alligator clips

Pickles

Ping pong balls

Black paint

Large balls

Internet

Textbooks

Different colored granular materials (examples: sand, sugar, dirt, flower, etc.)

Topographical maps

Geologic maps

Geologic time scales

Sugar cubes

Chalk

Vinegar

Pipettes

Aprons

Pitchers

Hairdryers

Spray bottles

Paste

5th Grade Science Investigations

Finding Properties Scavenger Hunt

Directions:

Use a **scale**, **graduated cylinder**, a **beaker of water**, and a **meter stick** to help you fill in the chart below, finding and measuring objects' properties.

Looking at the materials and lab we will be using, what safety precautions should we take to protect ourselves and the materials?

Property	Object	Measurement
A solid object		X
A liquid object		X
Magnetic object		X
Nonmagnetic object		X
An object that sinks in water		X
An object that floats in water		X
A gas		X
The mass of a solid object		
The mass of a liquid object		
The volume of a solid object		
The volume of a liquid		
The volume of a gas		

Separating Mixtures

Directions:

You will need a **magnet** in a **plastic baggy** (keep the magnet in the plastic baggy the whole time), a **wire strainer**, **filter paper**, a **hotplate**, **water**, **sand**, **sugar**, **marbles**, **iron filings**, **granola**, and two **beakers**.

Looking at the materials and lab we will be using, what safety precautions should we take to protect ourselves and materials during the investigation?

1) Take the mixture of sand, sugar, marbles, iron filings, and granola and use the materials to separate the mixture into its pieces. How will you divide the marbles from the mixture?

2) How will you separate the iron filings from the mixture?

3) How will you separate the granola from the mixture?

4) How will you separate the sand from the mixture?

5) How will you separate the sugar from the mixture?

Separating Pigments

Directions:

You will need **goggles**, **scissors**, different **pens** and **markers**, an **eyedropper**, **nail polish remover** or **alcohol**, **filter paper**, **test tubes**, a **test tube rack**, **paper clips**, and **rubber stoppers** with holes in them that fit the test tubes.

Looking at the materials and lab we will be using, what safety precautions should we take to protect ourselves and materials during the investigation?

1) You are going to do chromatography today. Cut the filter paper into strips and a point at one end.
2) Make a dot with the pen or marker about an inch from the pointy tip.
3) Take a paper clip and bend it, so there is a hook on one end that you poke through the flat end of the strip of filter paper and put the other end through the hole in the stopper.
4) Take the eyedropper and carefully squirt a little nail polish remover/alcohol into the bottom of a test tube. Then lower the paper with the dot pointy end down into the solvent, but do not let the dot touch it. Fix the stopper at the top of the test tube. Bend the part of the paper clip that is above the stopper so that the paper does not drop any lower.
5) Repeat the process you did for #s 1-4 for all the different pens and markers you have chosen.
6) Watch the patterns of pigments that separate. The pigments that move the fastest go to the top, and those that move the slowest will be at the bottom.
7) After your pigments have separated, remove them from the test tubes and let them dry (otherwise, all the stains may go back together again at the top of the paper).
8) Draw pictures of the color bands that came out from your pens and markers on your filter paper strips below:

Questions:

1) How did you see the same colors make different patterns?

2) Can the same color of ink be made of different substances?

   a. How did you see this today?

3) How could you use this information to determine which pen wrote a note if you need to?

Physical and Chemical Changes

Directions:

You will need **safety goggles**, an **aluminum pan**, **paper**, a **pipette**, a **candle**, a **lighter** or **matches**, **salt**, a **beaker** of **water**, **vinegar**, **baking soda** in a **small beaker**, **steel wool**, **long forceps**, a **clean piece of metal** and **another that is corroded** (of the same kind of metal), **hydrogen peroxide**, and **liver** or **banana**.

Looking at the materials and lab we will be using, what safety precautions should we take to protect ourselves and materials during the investigation?

1) **Physical changes** are changes that happen to a substance that does not cause any new substances to form. **Chemical changes** are changes that occur when new substances are formed.
2) Make sure everyone is wearing their protective goggles. Have your teacher come by and light the candle. What do you see happening to the wick of the candle?
3) Is this creating any new substances? Fill in Data Table 1 for the burning wick.
4) What do you see happening to the wax of the candle?
    a. Is this creating any new substances? Fill in Data Table 1 for the melting wax.
5) Take a piece of paper and tear it up. What did you see happen to the paper?

6) Did tearing the paper create any new substances? Fill in Data Table 1 for tearing paper.

7) Wad up a piece of paper and put it in the aluminum pan. Make sure everyone is wearing their safety goggles. Have your teacher come by with the lighter and light the paper on fire. Make sure to keep everything away from the flames. What do you see happening to the paper?

    a. Is it creating any new substances? Fill in Data Table 1 for burning paper.

8) Pour the salt into a beaker of water and stir it up. What do you see?

    a. Did this create any new substances? Fill in Data Table 1 for dissolving salt.

9) Now pour some vinegar into the small beaker of baking soda. What happened?

    a. Did this make any new substances? Fill in Data Table 1 for mixing vinegar and baking soda.

10) Look at the two pieces of metal. What difference do you see between the clean metal and the corroded?

11) Did corrosion cause a change to create a new substance? Fill in Data Table 1 for the metal.

12) Make sure you are wearing your safety goggles. Look at your steel wool and hold it with your forceps. Now have your teacher expose it to fire with the lighter. What do you see happen? Look at the color when your teacher is done (only the part of it that was lit).

    a. Did this produce any new substances? Fill in Data Table 1 for burning steel wool.

13) Finally, take some hydrogen peroxide and pour it on some liver or a slice of banana. What do you see happen?

    a. Did this produce a new substance? Fill in Data Table 1 below for hydrogen peroxide on life.

Data Table 1

Change that Happened	Physical or Chemical Change?	Evidence Why
Burning wick		
Melting wax		
Tearing Paper		
Burning paper		
Dissolving salt		
Vinegar and baking soda		
Corrosion of metal		
Burning of steel wool		
Hydrogen peroxide on life		

Questions:

1) Describe observations you might see when a physical change occurs.

2) Describe observations you might see when a chemical change occurs.

3) How could you tell dissolving sugar in water is a physical change?

4) Why does bubbling let you know there is a chemical change?

5) How do you know burning something is a chemical change?

Air Puck Motion

Directions:

You will need an **air puck**, five **stopwatches**, and **masking tape** to mark each meter with a **meter stick** on the floor for 5 meters.

Looking at the materials and lab we will be using, what safety precautions should we take to protect ourselves and materials during the investigation?

1) Have five students line up one meter apart, each with a stopwatch.
2) Everyone starts the stopwatch when the air puck moves past the starting line.
3) When the puck passes each person at their meter line, they must stop their stopwatch.
4) Record the times in Data Table 1 below.
5) Clear the stopwatches and repeat the procedure in #s 2-4. Then calculate the averages for each distance by adding the two times and dividing by two.
6) Graph the averaged data in the Time vs. Distance graph on the next page.
7) Find the graph's slope by taking the rise (distance) divided by the run (time).

Data Table 1

Distance (m)	Trial 1 Time (s)	Trial 2 Time (s)	Average Time (s)
0 m	0 s	0 s	0 s
1 m			
2 m			
3 m			
4 m			
5 m			

Graph 1

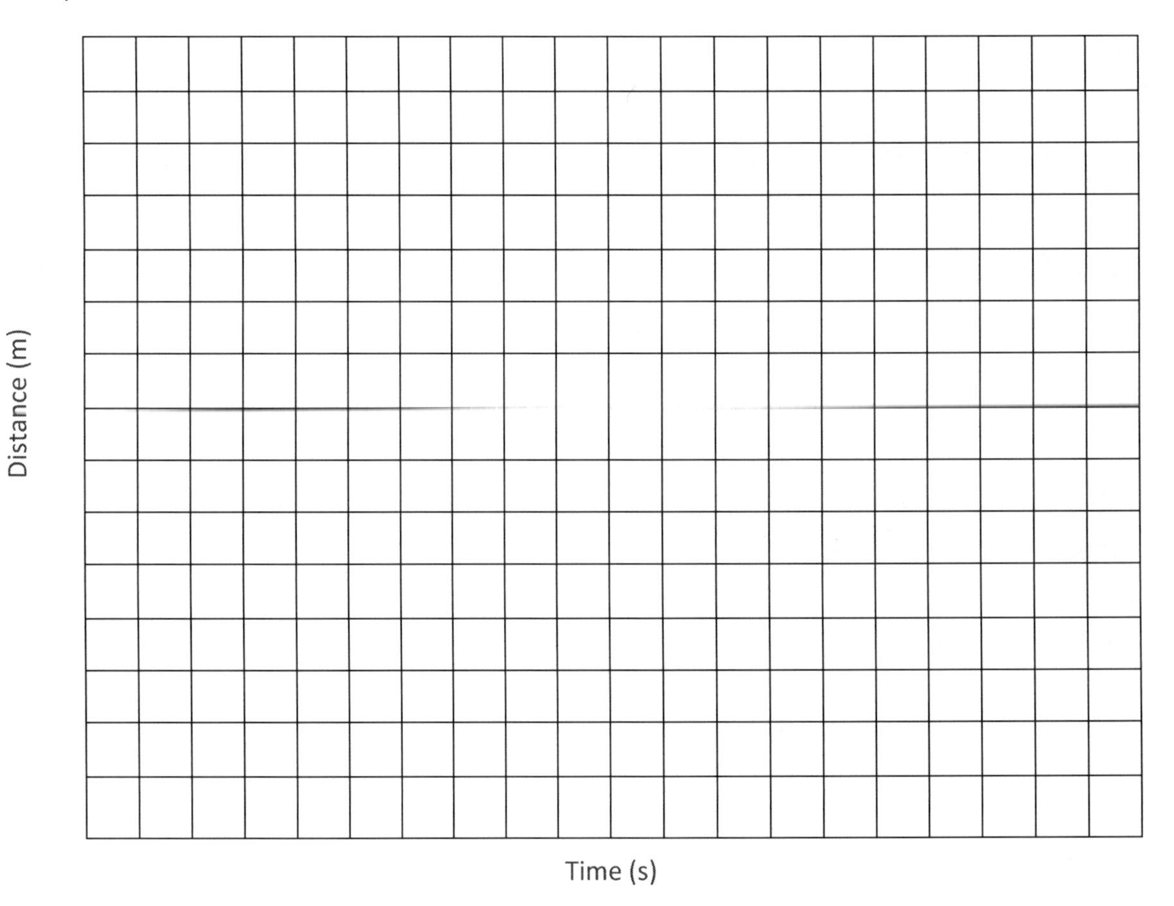

Questions:

1) On the distance-time graph, what does the slope of the line graph tell you?

2) What will a flat/horizontal line tell you on a distance-time graph?

3) What was the average speed of the air puck?

4) Did the speed of the puck seem constant or changing?

5) How far did the air puck travel while it was being timed?

Marbles in Motion

Directions:

Get a segment of **hot wheels track**, **small stickers**, a **marble,** and a **stopwatch**. Have something that one side of the track can be placed on to raise that end to create a ramp.

Looking at the materials and lab we will be using, what safety precautions should we take to protect ourselves and materials during the investigation?

1) Set the ramp so the bottom is along the edge of a tile on the floor. Most tiles in schools are 1 foot in length. Clear a path for 5 feet.
2) Adjust the height of the ramp so that the marble will just make it past 5 feet.
3) Place small stickers on the floor at the ramp's base and each foot past the base. The last one is 5 feet away from the base of the ramp.
4) Place a small sticker on the ramp to mark where you will place your marble for each trial to let it roll down the ramp; this keeps the distance your marble will be accelerating down the ramp constant.
5) Place your marble on the ramp and let it roll down (do not push). Time with a stopwatch how long it takes for the marble to move from the ramp's base to one foot away.
6) Repeat #5 four more times. Record the data in Data Table 1 below.
7) Repeat #s 5 and 6 for the distances 2 feet, 3 feet, 4 feet, and 5 feet away.
8) Find the average time for each distance.
9) Then calculate the average speed for each distance by taking the distance and dividing it by the average time and write that in Data Table 1.

Data Table 1

Trial	1 foot	2 feet	3 feet	4 feet	5 feet
1					
2					
3					
4					
5					
Average Time					
Average Speed					

10) Take the average speed and plot them on the graph to make a speed-distance graph; this will look similar to a velocity-time graph since the longer the distance, the more time it takes. The shape we see will be the same as looking at accelerated motion on a velocity-time graph.

Velocity vs. Distance Graph

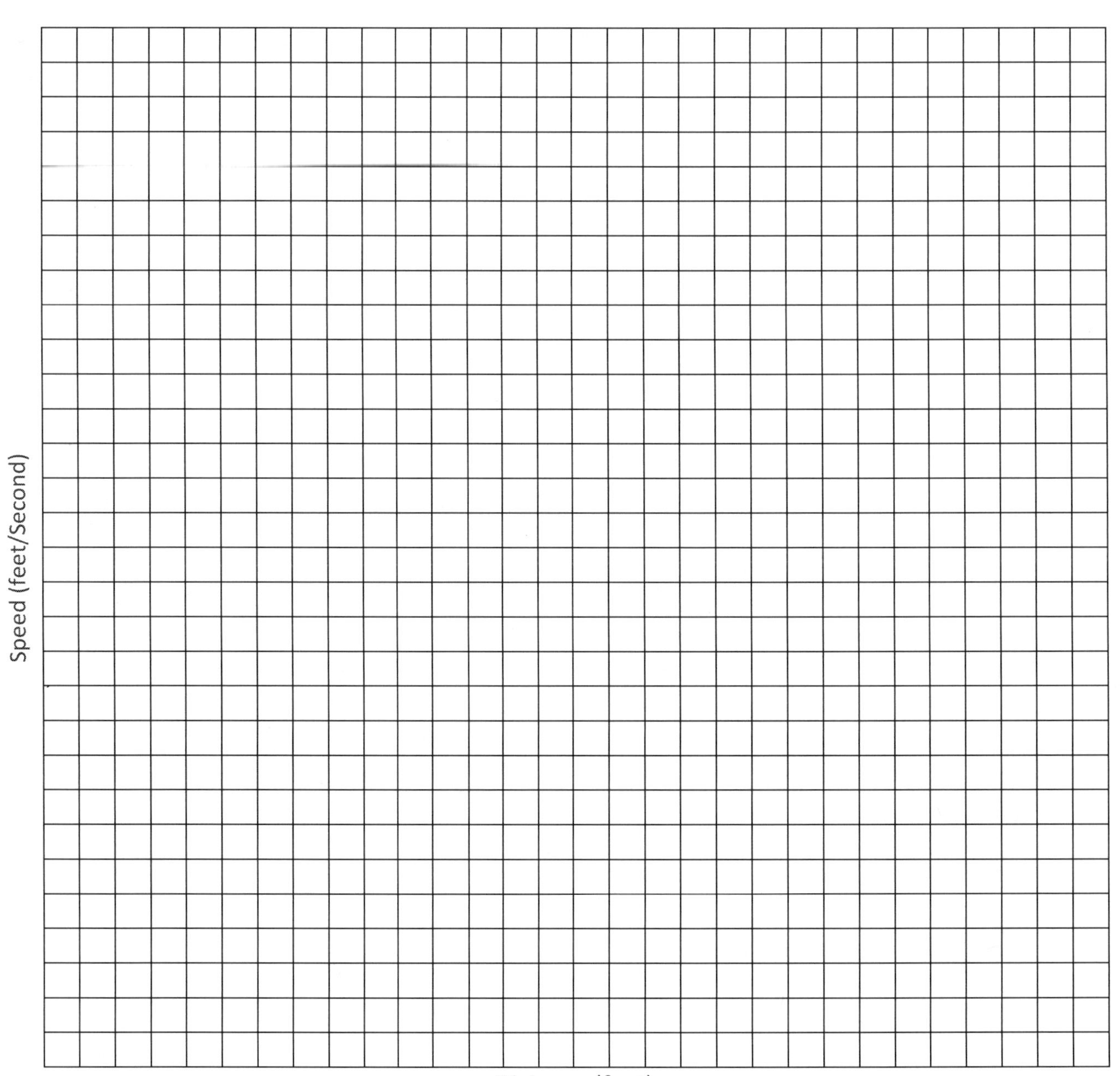

Questions:

1) Describe the motion of the marble as it moved down the ramp. If you have to, place the marble on the ramp and let it go to observe it move down the ramp.

2) What force caused the marble to speed up on the ramp?

    a. What is the direction of that force?

3) How could we make the marble have a higher velocity at the bottom of the ramp?

4) Describe the motion of the marble as it moved across the floor. Place the marble on the ramp again and observe it roll across the floor if you have to.

5) What caused the marble to slow down across the floor?

    a. What is the direction of that force?

6) How could we make the marble slow down faster across the floor?

7) At what distance would the average velocity be happing? At this distance would be the instantaneous velocity that is the same as the average velocity.

8) What was the shape of this graph? This graph shape is what acceleration looks like on a velocity-time graph.

Rocket Balloons

Directions:

You will need some **fishing lines** strung across the room, with a **straw** on each line. Inflate different types of **balloons** and tape them to the straw with the nozzle facing back. Have the students guess which ones will win the race across the room. Release the balloons and see which ones win. Use this setup to help you answer the following questions.

1) Which shape seems to travel the fastest?

2) Which shape seems to travel the farthest?

3) Which should win the race, a fully inflated balloon or one that is half inflated? Explain why.

4) Try it and explain the results.

5) What is causing the balloon to be forced forward?

6) What force caused the balloons to move slower?

7) What force caused the balloons to stop?

Build a Design to Win the Balloon Rocket Race

Directions:

Have the students use what they have learned to make modifications to make a balloon rocket that will win every race. Have the students race their balloon rockets to see who wins.

Too Small to be Seen

Directions:

Show some examples that particles are too small to be seen with the naked eye. However, have ways to detect them and show they can move into and out of objects.

1) How could we use a **balloon** to show unseen air can move into and out of a balloon?

2) How could we use a **small marshmallow** and a **large syringe** to show that air is inside a syringe?

3) If a **basketball** does not bounce, what could we do with an **air pump** and **needle** to make it bounce and show we added air into the basketball?

Conservation of Mass in a Phase Change

Directions:

Measure the mass of ice inside a Ziploc bag, let it melt, and measure the mass again. Write your measurements in Data Table 1.

Data Table 1

Mass of Ice in Bag	Mass of water in Bag

1) How does this show mass is conserved when a solid is changed into a liquid?

Energy Transformation

Directions:

Observe then explain how energy is transferred through the objects below. Explain what kind of energy is going in and how the energy is transformed into other forms until the object stops functioning.

1) **Plastic jumping frogs**

2) **Flashlight**

3) **Fireworks/bottle rockets**

4) **Ball** bouncing

5) **Toy car** rolling down a ramp of a **track**

Egg Drop

Directions:

Use any objects you have at home and/or anything your teacher gives you to build a device that absorbs the kinetic energy of impact when a **raw egg** is dropped from a height to be determined by your teacher. Draw a diagram of your design below showing how energy will be absorbed.

Your Design:

A Simple Circuit

Directions:

Attach the wires so that one end of a **battery pack** wire is connected to one end of the **Christmas light**. Take the other end of the Christmas light and attach it to one end of the **electric motor**. Take the other end of the motor and connect it to the battery pack to complete the circuit.

Looking at the materials and lab we will be using, what safety precautions should we take to protect ourselves and materials during the investigation?

1) Place **batteries** in the battery pack. What do you notice in the circuit that tells you it is closed and has an electric current?

2) Disconnect any wire; what happens?

   a. Is there a current flowing? How can you tell?

3) Cut out a little propeller and place it over the rod sticking out of the motor. Connect the wires again; in which direction does the motor spin?

4) Disconnect the battery pack and flip how the wires are connected to the circuit. How does the motor change?

   a. What do you think this tells you about the direction the electric current is flowing?

Human Circuits

Directions:

You will need a **Current Conductor** (found at Hobby Lobby) to show a circuit is closed by lighting up and making noise.

Looking at the materials and lab we will be using, what safety precautions should we take to protect ourselves and materials during the investigation?

Series Circuit:

1) Have students make a circle and hold hands while two of them hold the Current Conductor between them on the metal strips. When everyone touches hands in the circle, the circuit will be closed, and the Current Conductor will light up and make noise.
2) Open the Circuit: Have someone in the circle let go of someone's hand. What happened to the current of electricity?
3) Have everyone hold hands again, and at a different place, have someone let go. What happened?
4) Do you think this will happen every time in a series circuit? Explain why.

Parallel Circuit:

5) Have a few students come out of the circle and make one or two lines (depending on how big your class is) connecting across the inside of the circle. Make sure everyone is still touching hands. How do we know the circuit is closed?
6) Have students on the circle opposite the Current Conductor let go. What happened to the light and sound?

a. Why do you think this happened?

7) Where do people have to let go to make the circuit open? Experiment and find out.

8) If you had appliances in your house hooked up in a series circuit, what would happen if one of them was turned off?

9) How is a parallel circuit different than a series circuit?

10) Do you think your house is wired in series or parallel? Explain why.

11) How does this investigation show how energy is being transferred?

The Bending of Light

Directions:

Use a **lamp**, **mirror**, and a **hand lens** to find evidence of the behavior of light in the questions below.

1) Turn on the lamp and turn off the lights in the room. How does your shadow show light travels in a straight line?

2) Use the mirror to give evidence light is reflected. How does the mirror show you that light is reflected?

    a. Make the light of the lamp shine somewhere else by reflecting it with the mirror.

3) How does your shadow show light is absorbed?

4) Turn on the lights in the room and turn off the lamp. Look through the hand lens. How does it show you that light is bent (refracted)?

    a. How can bending light be useful?

Modeling Day and Night

Directions:

Modeling a Day:

Use a **globe** with a **lamp** to show how a 24-hour cycle happens on Earth. The lamp represents the sun shining on the Earth. Place a mark on the globe where you are located. Turn the globe counterclockwise to show the rotation of the Earth at different times of the day.

1) Show sunrise
2) Show noon
3) Show sunset
4) Show midnight
5) Why does the sun appear to move across the sky each day?

Modeling the Phases of the Moon:

Now take a **baseball** to represent the moon and place it where it will be in relation to the Earth and sun for the different phases below.

1) Show new moon
2) Show the first quarter
3) Show full moon
4) Show last quarter
5) Why does the moon appear to change shape each month?
6) When is the moon in the sky during the day?

7) Does the moon...?
    a. Spend more time in the sky at night
    b. Spend more time in the sky during the day
    c. Spend the same amount of time in the sky during the day as at night

8) Why do other stars not have the same impact on Earth with their light as our sun?

Building a Sundial

Directions:

Take a **wooden dowel** and stick it in the ground at 9:00 am, tilting it slightly north. Place a **rock** down where the shadow is cast. At 10:00 am, go out and place a **rock** where the shadow is cast. Keep going outside each hour and place a **rock** down where the shadow is each hour until school is out. The teacher can stay late and come in early to complete all daylight hours on the sundial.

1) What is the source of light?

2) What is casting a shadow?

3) How does the shadow move during the day?

4) How does the sundial help us tell time during the day?

5) Over time, do you think the shadows will be cast in the same places every day?

6) Wait a month and check to see if the shadows shift at any of the times. What did you find?

The Water Cycle

Directions:

Use the **internet** or your **textbook** to draw the water cycle below. Pay close attention to what directions the water moves to make it condense into clouds and rain.

Questions:

1) What direction does water have to move to condense into clouds and rain?

    a. What happens to storms if forces make air move faster in that direction?

How Hurricanes Form

Directions:

Use the **internet** to go to a NASA URL address at https://tinyurl.com/4erbxnet. Use this web page to research how hurricanes form and answer the questions below.

1) Where do tropical cyclones form on Earth?

    a. Why do they form there?

2) What do hurricanes need to form?

3) Describe how a hurricane forms.

4) How does the eye form?

5) Why do hurricanes weaken when they go over land?

6) What determines the category of the hurricane?

7) Describe each category of hurricane listed here.

   a. 1

   b. 2

   c. 3

   d. 4

   e. 5

8) What causes the damage we see on land?

9) What allows us to better forecast where hurricanes will hit?

10) How do forecasters show where they predict the hurricanes will move to?

11) What can cause an increase in the number and intensity of hurricanes?

Sedimentary Rock Model

Directions:

Either cook different colors of **cookies** or buy different colors of cookies. Take the cookies (dry and crumbly) and place the same colors in the same **Ziploc bag**. Each color will have a different bag. Break up the cookies in the bags to make the sediment. Mix some **honey** and **water** together and make sure each group has some in a **small beaker** with a **pipette**. Get some **small cups** and **large sprinkles**. Follow the directions below to model how sedimentary rock forms.

1) Take your cup and add a color of cookie crumbs and make a layer on the bottom of the cup.
2) Add honey water over the top with the pipette to make it moist but not saturated.
3) Take a different color of cookie crumbs and as you add this layer, add one color of sprinkles to be organisms that died and are now buried in this layer.
4) Add honey water over the top with the pipette to make it moist but not saturated.
5) Take a different color of cookie crumbs from the last and add another layer while putting in some sprinkles of a different color in this layer.
6) Add honey water over the top with the pipette to make it moist but not saturated.
7) Push down to simulate the sediment's compaction when buried to produce the rock.
8) Keep adding different color layers in the cup, sprinkles, and honey water until your cup is full.
9) Push down to compact it one more time.
10) Flip the cup over and have the contents come out.
11) Flip your sedimentary rock model right side up and answer the following questions.

Questions:

1) Which layer was the first to be laid down?

2) Which layer is the oldest?

3) Which layer was the last to be laid down?

4) Which layer is the youngest?

5) Where are the oldest fossils?

6) Where are the youngest fossils?

7) Cut a section off of your sedimentary rock formation. How are you able to find the fossils (sprinkles)?

    a. Did any of the fossil layers mix?

8) If enough organisms die and are buried together as they decompose underground away from the air, they turn into fossil fuels. How could you change this model to have it model the trapping and production of fossil fuels?

9) Nature has been burying fossils underground for millions of years, trapping these chemicals underground and causing the atmosphere to become thinner over time. What do you think will happen to our atmosphere if we dig them up and burn them back into the atmosphere?

Sand Dune, Canyon, and Delta Formations

Directions:

Put on some **safety goggles**, pour some **dry sand** into a **large tub**, and plug in a **hair dryer**.

1) Turn on the hair dryer and point it at the sand. What do you notice happening to the sand?

2) Describe how sand dunes are made in a dry desert.

3) Raise one end of your tub and pour a **beaker of water** into the high end of the tub on the sand. What do you notice happening to the sand?

4) Fill the beaker again and pour in more water. The more water you pour into the same spot, what happens to the size of this formation?

5) Describe how canyons are formed.

6) What do you notice building up at the lower end of the tub?

7) How do you think deltas form?

Conservation of Resources Project

Directions:

Choose one of the natural resources below, research how it is used, and develop a way to conserve this resource, so we do not run out and do no harm to the environment. Present this information in the format of your teacher's choosing.

Water

Coal

Petroleum

Wood

Uranium

Natural Gas

Metal ores

Fruit and vegetables

Wind

Hydroelectricity

Solar Energy

Geothermal Energy

Meat (like fish, beef, and chicken)

Cement

Land

Cookie Cooker

Directions:

In an effort to conserve resources while cooking food, use materials at home and/or materials your teachers provide you to build a solar cookie cooker. Students will build a device to cook **cookie dough** completely in the shortest amount of time. You can use live demonstration or video evidence of the time it took to completely cook a cookie to your teacher's specifications. Draw your final design below of your device. Compare the designs of the fastest cookers to see what they have in common.

Your Design:

The time it took to cook: ____

What qualities do the fastest cookers have in common?

Biotic and Abiotic Factors

Directions:

You will need a piece of **paper**, a **clear sheet protector**, and some **markers**.

1) On the piece of paper, draw the abiotic (nonliving parts of an ecosystem) factors in a healthy ecosystem. Examples: rocks, ground, sun, clouds, earth formations, lakes, or rivers. Place the word Abiotic at the top of the page.
2) Place the paper you drew on inside the clear sheet protector. On the clear sheet protector, draw the biotic (living parts of the ecosystem) factors as they would be in a healthy ecosystem. Examples: plants, animals, and fungus. Place the word Biotic at the bottom of the page.
3) Together you have an ecosystem. Whenever you want to differentiate between the Biotic and Abiotic parts of the ecosystem, just take the paper out of the clear sheet protector.

Questions:

1) If you were to run out of water in your ecosystem, what would happen to it?

2) If your carnivores were killed off in your ecosystem, how would that affect your herbivores?

    a. How would that affect the plants?

3) What would happen to your ecosystem if humans moved in and built buildings and roads?

Pros and Cons of Human Activities

Directions:

On the left side of the chart below, write human activities that benefit an ecosystem. On the right-hand side, write human activities that harm an ecosystem. Discuss with your class what you notice.

Beneficial Human Activities	Harmful Human Activities

What allows these animals to survive underwater?

Directions:

Look at the pictures of the aquatic animals below and under each tell how their structures allow them to survive underwater.

1) 2) 3)

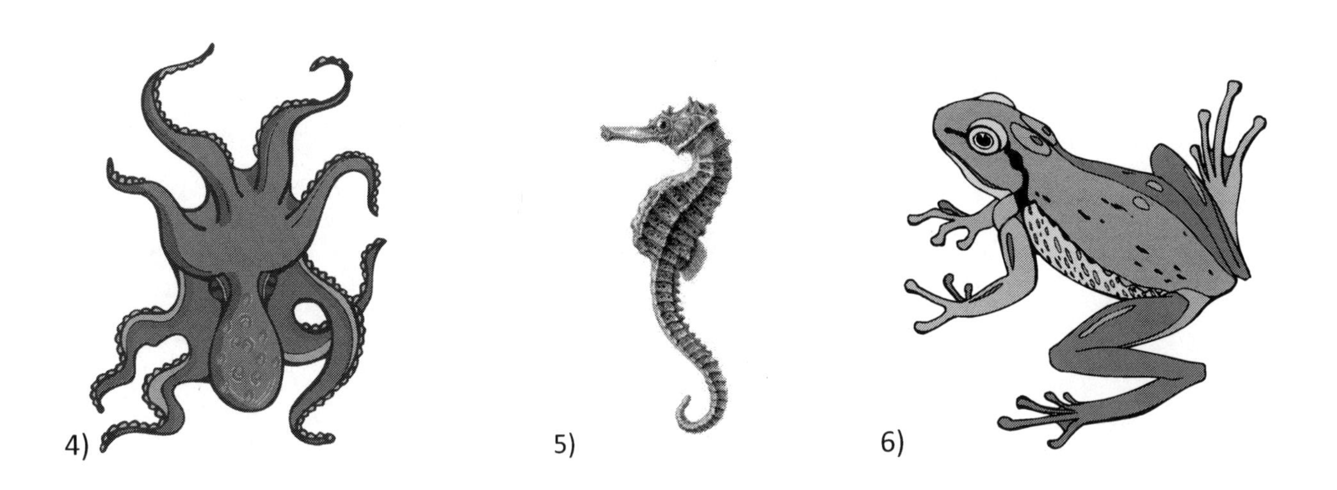

4) 5) 6)

What allows these organisms to survive in a desert?

Directions:

Look at the pictures of the organisms below, research each, and under each, tell how their structures allow them to survive in a desert.

1) 

2) 

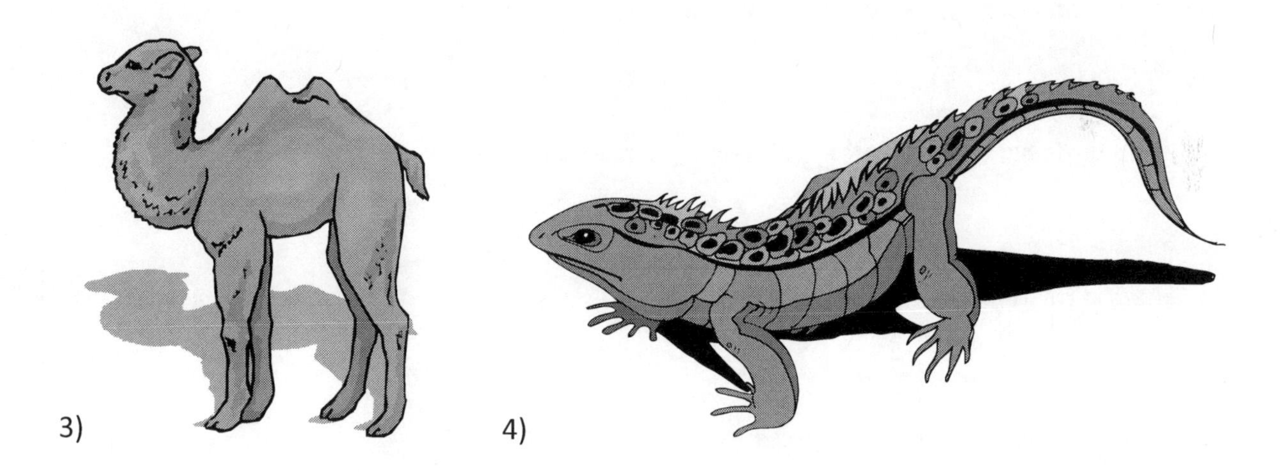

3)

4)

Instinctual and Learned Behavior

Directions:

Organisms are born with **instinctual behavior** that all members of their species perform. **Learned behavior** only occurs after experience or practice. Determine if each of these behaviors is instinctual or learned, then tell how this behavior increases the animal's chances of survival.

1) Monarch migration (**instinct** or **learned**):

2) Cat using a litterbox (**instinct** or **learned**):

3) Tiger hunting and killing prey (**instinct** or **learned**):

4) Selfishness (**instinct** or **learned**):

5) Turtle hatchling returning to the sea (**instinct** or **learned**):

6) Male birds fluffing up feathers in front of females (**instinct** or **learned**):

7) Sharing (**instinct** or **learned**):

8) Salmon swimming upstream to spawn (**instinct** or **learned**):

9) Orcas hunt in packs (**instinct** or **learned**):

10) Love (**instinct** or **learned**):

Photosynthesis and Plant Anatomy

Directions:

The formulas below show plants take in sunlight, water, and carbon dioxide to make food and oxygen. Draw a picture of a plant below and show where the sunlight, water, and carbon dioxide go into the plant, where the food is stored (the parts animals eat), and where the oxygen comes out. Use the **internet** and/or your **textbook** to research how these things enter and leave a plant.

Water + carbon dioxide + sunlight ⟶ food + oxygen

$$H_2O + CO_2 + \text{sunlight} \longrightarrow C_6H_{12}O_6 + O_2$$

Food Webs and Pyramids

Directions:

Use the diagram below to help answer the questions.

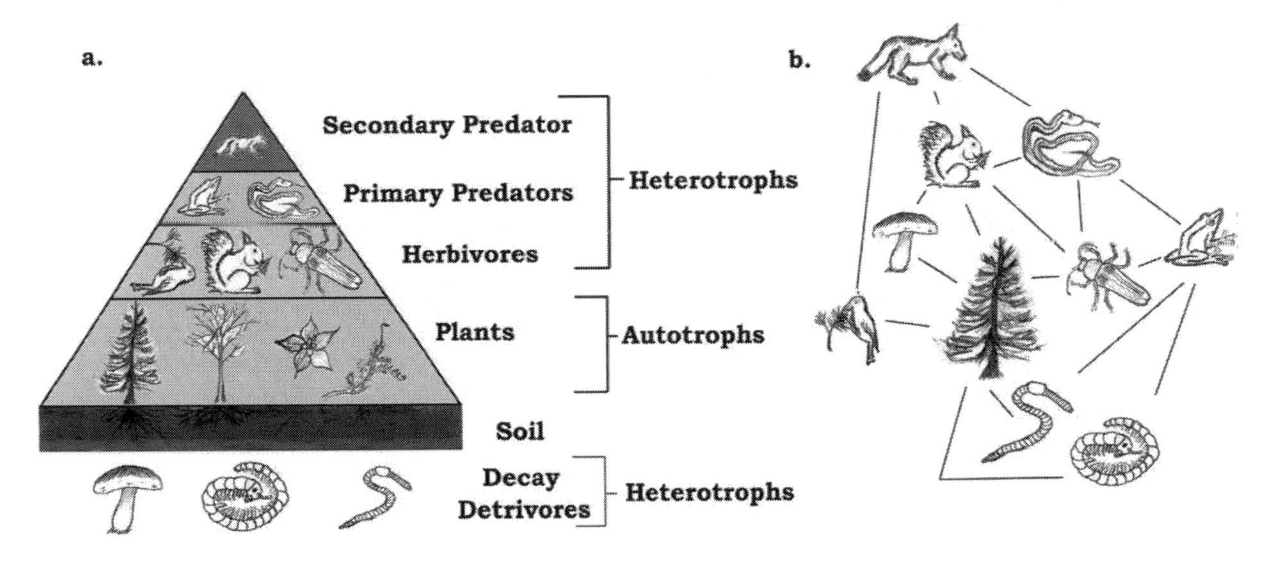

Diagram from Thompsma

1) Put arrowheads on the lines in diagram (b) to show who eats who.
2) What role does the sun play in the community?

   a. What would happen to the community if the sun were to go away?

3) What role do the plants play in this community?

   b. What would happen to this community if the plants were to go away?

4) What role do the Herbivores play in the community?

c. What would happen to this community if the herbivores were to go away?

5) What role do the predators play in the community?

d. What would happen to this community if the predators were to go away?

6) What role do the decomposers (decayers/detrivores) play in the community?

e. What would happen to the community if the decomposers were to go away?

7) How do all of these organisms depend on each other for their survival?

8) What is the ultimate energy source of the material that makes up a carnivore in this or any ecosystem on Earth?

9) What are typical ways humans could disrupt this community?

10) What do you think interdependence means?

Managing Garden Soil Moisture

Directions:

You will need a **heat lamp**, two shoebox-sized **tubs** of **soil** from the same source, one with some **mulch** covering the top and one without mulch. You will also need a **soil moisture probe** attached to an **interface** connected to a **computer** with **Logger Pro**.

Looking at the materials and lab we will be using, what safety precautions should we take to protect ourselves and materials during the investigation?

1) Pour 500 mL of water evenly into each of the boxes of soil. Let it sit for a few minutes and measure the amount of soil moisture in each box by placing the soil moisture probe into the soil where the two prongs are vertical to each other and carefully pushing the probe into the soil until it is entirely under the soil. Write this data in Data Table 1.
2) Let the soil samples sit under a heat lamp for three-four days.
3) Then measure the moisture using the soil moisture probe again for each box of soil. Write this data in Data Table 1.

Data Table 1

	Soil With Mulch	Soil Without Mulch
Moisture Before		
Moisture After		
Change in Moisture		

Questions:

1) Which sample of soil held more moisture in it?

2) Which sample would you want to be set up for your bushes and flowers at your house? Explain Why.

3) What is the main purpose of mulch?

4) Which setup would allow you to use less water to keep your plants alive?

5) What are three materials in our area that are used for mulch?

Organic Gardening and Hydroponics

Directions:

Use the **internet** to research Organic Gardening and Hydroponics to help you answer the following questions.

1) What is organic gardening?

2) What are some examples of organic gardening?

3) What are the benefits of organic gardening?

4) What are the negative aspects of organic gardening?

5) What is hydroponics?

6) What are some examples of hydroponics?

7) What are the benefits of hydroponics?

8) What are the negative aspects of hydroponics?

Where is our Water?

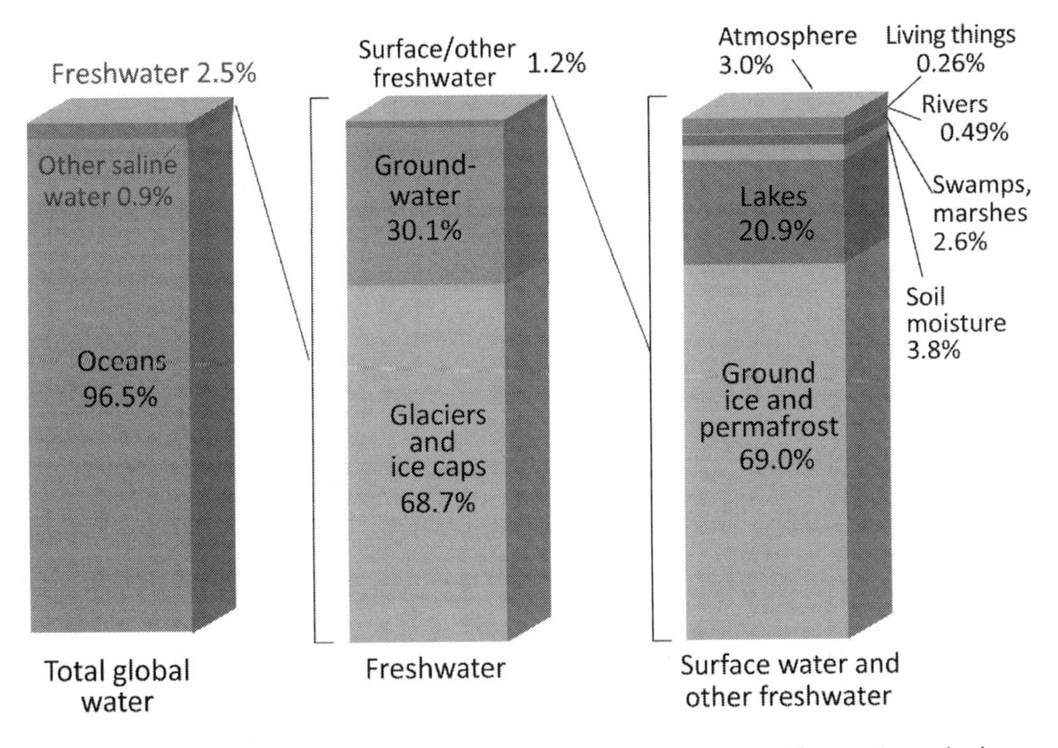

Credit: U.S. Geological Survey, Water Science School. https://www.usgs.gov/special-topic/water-science-school
Data source: Igor Shiklomanov's chapter "World fresh water resources" in Peter H. Gleick (editor), 1993, Water in Crisis: A Guide to the World's Fresh Water Resources. (Numbers are rounded).

Directions:

Use the image above to help you answer the following questions.

1) How much of the total water on Earth is freshwater?
2) How much of that is locked up in ice?
3) How much freshwater is usable and renewable (on the surface)?
4) How much is groundwater and not renewable?
5) With your teacher and class, calculate how much of the total global water is renewable freshwater in liquid form.

Virtual Investigations that go with 5th Grade Science

ExploreLearning.com

Density

Density Experiment: Slice and Dice

Measuring Volume

Mass and Weight

Circuit Builder

Mineral Identification

Phases of Water

Force and Fan Carts

Conduction and Convection

Radiation

Pendulum Clock

Energy Conversions

Incline Plane Rolling Objects

Basic Prism

Carbon Cycle

Rock Cycle

River Erosion

Weathering

Erosion Rates

Hurricane Motion

Comparing Climates (Customary)

Comparing Climates (Metric)

Building Topographical Maps

Weather Maps

Weather Maps – Metric

Eclipse

Comparing Earth and Venus

Seasons Earth, Moon, and Sun

Phases of the Moon

Pond Ecosystem

Forest Ecosystem

Ecosystems – Undefined STEM Case

Inheritance

Effects of Environment on New Life Form

Observing Weather (Customary)

Observing Weather (Metric)

Water Cycle

Prairie Ecosystem

Rabbit Population by Season

Growing Plants

Seed Germination

Eyes and Vision 1

Photosynthesis Lab

Cell Energy Cycle

Programmable Rover

PhET.colorado.edu

Density

Gases Intro

States of Matter: Basics

Concentration

Projectile Motion

Pendulum Lab

Energy Forms and Change

Gravity Force: Basics

Energy Skate Park: Basics

Balancing Act

Friction

Forces and Motion: Basics

States of Matter: Basics

Bending Light

Circuit Construction Kits: DC – Virtual Lab

Circuit Construction Kits: DC

John Travoltage

Balloons and Static Electricity

Gravity and Orbits

Natural Selection

Physicsclassroom.com/Physics-Interactives

Force

Rocket Sledder

Skydiving

Egg Drop

Kinetic Energy

It's All Uphill

Orbital Motion

Gravitation

DC Circuit Builder

Plane Mirror Images

Refraction and Lenses

5th Grade Science TEKS and NGSS Correlations

Finding Properties Scavenger Hunt Science, 5th Grade (b) 1ABCDEF 5C 6A; 5-PS1-3

Separating Mixtures Science, 5th Grade (b) 1ABCDE 3AB 5ABG 6BC

Separating Pigments Science, 5th Grade (b) 1ABCDE 3AB 5ABG 6BC

Physical and Chemical Changes Science, 5th Grade (b) 1ABCDEF 2B 3AB 5ABEG 6BC; 5-PS1-34

Air Puck Motion Science, 5th Grade (b) 1ABCDEF 2BC 3AB 5ABCDEG 7A

Marbles in Motion Science, 5th Grade (b) 1ABCDEF 2BC 3AB 5ABCDEG 7AB

Rocket Balloons Science, 5th Grade (b) 1ABCDE 2B 3AB 5ABCDEG 6A 7AB

Build a Design to Win a Balloon Rocket Race Science, 5th Grade (b) 1ABCDE 2BCD 4A 5ABCDEG 6A 7AB; 5-PS1-1, 3-5-ETS1-123

Too Small to be Seen Science, 5th Grade (b) 1ABCDEG 2ABC 3ABC 5ABCDEG 6AD; 5-PS1-1, 3-5-ETS1-1

Conservation of Mass in a Change Science, 5th Grade (b) 1ABCDEF 2BC 3AB 5ABCDG 6A; 5-PS1-2

Energy Transformation Science, 5th Grade (b) 1ABCDE 3ABC 5ABCDEFG 7A 8AB; 5-PS2-1

Egg Drop Science, 5th Grade (b) 1ABCDE 2BD 4A 5ABCDEFG 7AB 8A; 5-PS2-1, 3-5-ETS1-123

A Simple Circuit Science, 5th Grade (b) 1ABCDE 2B 3ABC 5ABCDEFG 6A 8AB

Human Circuits Science, 5th Grade (b) 1ABCDE 2BD 3ABC 5ABCDEFG 6A 8AB

The Bending of Light Science, 5th Grade (b) 1ABCDE 2B 3ABC 5ABCDE 8C

Modeling of Day and Night Science, 5th Grade (b) 1ABCDEG 2AB 3ABC 5ABCDEFG 8C 9; 5-ESS1-1

Building a Sundial Science, 5th Grade (b) 1ABCDEG 2ABD 3ABC 5ABCDEFG 8C 9; 5-ESS1-2

The Water Cycle Science, 5th Grade (b) 1ABEFG 3AB 5ABDEFG 10A

How Hurricanes Form Science, 5th Grade (b) 1ABE 3ABC 5ABDEFG 10A

Sedimentary Rock Model Science, 5th Grade (b) 1ABCD 2AB 3ABC 5ABCDG 10B; 5-ESS2-1

Sand Dune, Canyon, and Delta Formations Science, 5th Grade (b) 1ABCD 2AB 3ABC 5ABCDEG 10C; 5-ESS2-1

Conservation of Resources Project Science, 5th Grade (b) 1AB 3ABC 4AB 5ABDEF 11; 5-ESS2-2, 5-ESS3-1, 3-5-ETS1-1

Cookie Cooker Science, 5th Grade (b) 1ABCDEFG 2ABD 3ABC 4AB 5ABCDEFG 6A; 3-5-ETS1-123

Biotic and Abiotic Factors Science, 5th Grade (b) 1ABEFG 3AB 5ADEFG 12ABC

Pros and Cons of Human Activities Science, 5th Grade (b) 1ABF 5BDEFG 12C; 5-ESS3-1

What allows these animals to survive underwater? Science, 5th Grade (b) 1ABEG 2A 3AB 5ABF 13A

What allows these animals to survive in a desert? Science, 5th Grade (b) 1ABEG 2A 3AB 5ABF 13A

Instinctual and Learned Behavior Science, 5th Grade (b) 1AB 3AB 4B 5ABDEFG 13AB

Photosynthesis and Plant Anatomy Science, 5th Grade (b) 1ABFG 3AB 4B 5DEFG 13A; 5-LS1-1

Food Webs and Pyramids Science, 5th Grade (b) 1ABEG 3AB 5BDEFG 12BC; 5-PS3-1, 5-LS1-1, 5-LS2-1

Managing Garden Soil Moisture Science, 5th Grade (b) 1ABCDEF 3AB 5ABCDEFG 11 12AB; 5-PS1-3, 5-ESS3-1

Organic Gardening and Hydroponics Science, 5th Grade (b) 1ABE 3AB 4AB 5BDEFG 11; 5-ESS3-1

Where is our Water? Science, 5th Grade (b) 1ABEG 3ABC 5ACDE; 5-ESS2-2

Equipment List for all 5th Grade Investigations

If you want to be able to do all the labs in this manual for 5th Grade Science, here is a list of all the equipment you will need (in order of appearance).

Scales

Graduated cylinders

Beakers

Water

Meter sticks

Magnets

Plastic baggies

Wire strainers

Filter paper

Hotplates

Sand

Sugar

Marbles

Iron filings

Granola

Goggles

Scissors

Pens

Markers

Eye droppers

Nail polish remover

Alcohol

Test tubes

Test tube racks

Paper clips

Rubber stoppers

Aluminum pans

Paper

Pipettes

Candles

Lighter

Matches

Salt

Vinegar

Baking soda

Steel wool

Long forceps

A clean piece of metal

Corroded piece of metal

Hydrogen peroxide

Liver

Banana

Air puck

Stopwatches

Masking tape

Hot wheels track

Small stickers

Marbles

Fishing line

Straws

Balloons

Small marshmallows

Large syringe

Basketball

Air pump and needle

Plastic jumping frogs

Flashlights

Balls

Toy cars

Raw eggs

Battery packs

Christmas lights

Electric motors

Batteries

Current Conductor

Lamp

Mirrors

Hand lenses

Globe

Baseball

Wooden dowel

Rocks

Internet

Textbooks

Cookies of different color

Ziploc bags

Honey

Small cups

Large sprinkles

Safety goggles

Dry sand

Large tubs

Hair dryers

Raw cookie dough

Clear sheet protectors

Markers

Heat lamp

Soil

Mulch

Soil moisture probe

Interface

Logger Pro

Computer

Made in the USA
Columbia, SC
10 September 2024

41484344R00154